Risk

D1344987

Safety at Work Series*

Volume 1 – Safety Law

Volume 2 – Risk Management

Volume 3 – Occupational Health and Hygiene

Volume 4 – Workplace Safety

---

*These four volumes are available as a single volume, *Safety at Work*, 5th edition.

Risk Management

Volume 2 of the Safety at Work Series

Edited by
John Ridley and John Channing

OXFORD AUCKLAND BOSTON JOHANNESBURG MELBOURNE NEW DELHI

Butterworth-Heinemann
Linacre House, Jordan Hill, Oxford OX2 8DP
225 Wildwood Avenue, Woburn, MA 01801-2041
A division of Reed Educational and Professional Publishing Ltd

℞ A member of the Reed Elsevier plc group

First published 1999

British Library Cataloguing in Publication Data
A catalogue record for this book is available from the British Library

Library of Congress Cataloguing in Publication Data
A catalogue record for this book is available from the Library of Congress

ISBN 0 7506 4558 X

Composition by Genesis Typesetting, Laser Quay, Rochester, Kent
Printed and bound in Great Britain by
Biddles Ltd, Guildford and King's Lynn

Contents

Foreword

Frank J. Davies CBE, O St J, *Chairman, Health and Safety Commission*

My forty years experience of working in industry have taught me the importance of health and safety. Even so, since becoming Chairman of the Health and safety Commission (HSC) in October 1993, I have learned more about the extent to which health and safety issues impact upon so much of our economic activity. The humanitarian arguments for health and safety should be enough, but if they are not the economic ones are unanswerable now that health and safety costs British industry between £4 billion and £9 billion a year. Industry cannot afford to overlook these factors and needs to find a way of managing health and safety for its workers and for its businesses.

In his foreword to the third edition of this book my predecessor, Sir John Cullen, commented on the increasing impact of Europe in the field of health and safety, most notably through European Community Directives. We have since found this to be very much so. I believe that the key challenge health and safety now faces is to engage and influence the huge variety of businesses, particularly small businesses, and to help them manage health and safety more effectively. I would add that the public sectors, our largest employers these days, also should look at their management of health and safety to ensure they are doing enough.

Many businesses are willing to meet their legal obligations if given a gentle prompt and the right advice and HSC is very conscious of the importance of having good regulations which are practicable and achievable.

It is, of course, vital and inescapable that an issue as critical as health and safety should be grounded in sound and effective legislation.

This book covers many of these and other important health and safety developments, including environmental and industrial relations law which touch on this area to varying degrees. I welcome the contribution it makes towards the goal of reaching and maintaining effective health and safety policies and practices throughout the workplace.

Preface

Health and safety is not a subject in its own right but is an integration of knowledge and information from a wide spectrum of disciplines. *Safety at Work* reflects this in the range of chapters written by experts and in bringing the benefits of their specialised experiences and knowledge together in a single volume.

While there is a continuing demand for a single volume, many managers and safety practitioners enter the field of safety with some qualifications already gained in an earlier part of their career. Their need is to add to their store of knowledge specific information in a particular sector. Equally, new students of the subject may embark on a course of modular study spread over several years, studying one module at a time. Thus there appears to be a need for each part of *Safety at Work* to be available as a stand-alone volume.

We have met this need by making each part of *Safety at Work* into a separate volume whilst, at the same time, maintaining the cohesion of the complete work. This has required a revision of the presentation of the text and we have introduced a pagination system that is equally suitable for four separate volumes and for a single comprehensive tome. The numbering of pages, figures and tables has been designed so as to be identified with the particular volume but will, when the separate volumes are placed together as a single entity, provide a coherent pagination system.

Each volume, in addition to its contents list and list of contributors, has appendices that contain reference information to all four volumes. Thus the reader will not only have access to the detailed content of the particular volume but also information that will refer him to, and give him an overview of, the wider fields of health and safety that are covered in the other three volumes.

In this way we hope we have kept in perspective the fact that while each volume is a separate part, it is only one part, albeit a vital part, of a much wider spectrum of disciplines that go to make occupational health and safety.

John Ridley
John Channing
October 1998

Contributors

L. Bamber, BSc, DIS, FIRM, FIOSH, RSP
Principal Consultant, Norwich Union Risk Services
Dr A.J. Boyle, PhD, BSc, ABPhS
Health and Safety Technology and Management Ltd
John Channing, MSc(Chem), MSc(Safety), FIOSH, RSP
Manager Health, Safety and Environment, Kodak Manufacturing
Professor Andrew Hale, PhD, CPsychol., MErgS, FIOSH
Professor of General Safety Science, Delft University of Technology.

Introduction

In every activity there is an element of risk and the successful manager is the one who can look ahead, foresee the risks and eliminate or reduce their effects. Risks are no longer confined to the 'sharp end', the shop floor, but all parts of the organisation have roles to play in reducing or eliminating them. Indeed, the Robens' Committee recognised the vital role of management in engendering the right attitudes to, and developing high standards of, health and safety throughout the organisation.

A number of specialised techniques have been developed to enable risks to be identified, assessed and either avoided or reduced but there are other factors related to the culture of the organisation and the inter-relationship of those who inhabit it that have a significant role to play. An understanding of those techniques and the roles and responsibilities of individuals and groups is a necessary prerequisite for high levels of safety performance.

Chapter 1

Principles of the management of risk

L. Bamber

1.1 Principles of action necessary to prevent accidents

1.1.1 Introduction

The Ministry of Labour and National Service[1] postulated six principles of accident prevention in 1956 that are still valid today. These are:

1 Accident prevention is an essential part of good management and of good workmanship.
2 Management and workers must co-operate wholeheartedly in securing freedom from accidents.
3 Top management must take the lead in organising safety in the works.
4 There must be a definite and known safety policy in each workplace.
5 The organisation and resources necessary to carry out the policy must exist.
6 The best available knowledge and methods must be applied.

It would appear that these principles have only received legislative backing in more recent times – i.e. via the Health and Safety at Work etc. Act 1974, the Safety Representatives and Safety Committees Regulations 1977 and via the European Union, through the Management of Health and Safety at Work Regulations 1992 and the Health and Safety (Consultation with Employees) Regulations 1996.

Before a closer examination of principles of action necessary to prevent accidents is undertaken, there is a need to examine more closely what is meant by an accident.

1.1.2 What is an accident?

To start, consider the following axiomatic statements:

1 All accidents are incidents.
2 All incidents are *not* accidents.

3 All injuries result from accidents.
4 All accidents do *not* result in injury.

An early definition was propounded by Lord MacNaughton in the case of *Fenton* v. *Thorley & Co. Ltd* (1903) AC 443 where he defined an accident as 'some concrete happening which intervenes or obtrudes itself upon the normal course of employment. It has the ordinary everyday meaning of an unlooked-for mishap or an untoward event, which is not expected or designed by the victim.'

This definition refers to an event occurring to a worker that was an unlooked-for mishap having a degree of unexpectedness about it. However, taking into account the axiomatic statements above, this definition would seem to be somewhat narrow, as it is only concerned with accidents resulting in injury to employees.

From research of some 40 accident definitions from general, legal, medical, scientific and safety literature, it appears that the ideal accident definition should have two distinct sections: a description of the causes, and a description of the effects.

Causes should include: unexpectedness or unplanned events, multi-causality and sequence of events; while the effects should cover: injury, disease, damage, near-miss and loss.

Based on the research, the following definition is suggested: 'an accident is an unexpected, unplanned event in a sequence of events, that occurs through a combination of causes; it results in physical harm (injury or disease) to an individual, damage to property, a near-miss, a loss, or any combination of these effects'.

This definition requires recognition of a wider range of accidents than those resulting in injury.

1.2 Definitions of hazard, risk and danger

The HSE leaflet *Hazard and Risk Explained*[2] presents the definitions of 'hazard' and 'risk' in relation to the COSHH Regulations:

1.2.1 Hazard

The *hazard* presented by a substance is its potential to cause harm. Hazard is associated with degrees of danger, and is quantifiable.

1.2.2 Risk

The *risk* from a substance is the likelihood that it will cause harm in the actual circumstances of use. This will depend on: the hazard presented by the substance; how it is used; how it is controlled; who is exposed . . . to how much . . . for how long. Risk should be thought of in terms of 'chance-taking'. What are the odds – the probability – of an accident occurring? Risk can be taken after careful consideration, or out of ignorance. The result can be fortuitous or disastrous, or anything in between.

The link between 'hazard' and 'risk' must be understood. In terms of the COSHH Regulations, poor control can create substantial *risk* even from a substance with low *hazard*. But with proper controls, the risk of being harmed by even the most hazardous substance is greatly reduced.

Also, the Approved Code of Practice[3] for the Management of Health and Safety at Work Regulations 1992 clearly states:

(a) a hazard is something with the potential to cause harm (this can include substances or machines, methods of work and other aspects of work organisation);
(b) risk expresses the likelihood that the harm from a particular hazard is realised;
(c) the extent of the risk covers the population which might be affected by a risk; i.e. the number of people who might be exposed and the consequences for them.

Risk, therefore, reflects both the likelihood that harm will occur and its severity. Hence these factors should be taken into account when undertaking either qualitative or quantitative risk assessment.

Danger can be associated with situations where there is a distinct possibility of:

(a) Interchanges of energy above tolerance levels. Such interchanges can be equated to any form of matter, including animals, vegetables and inert objects. The interchanges in energy can be in the form of physical, chemical, biological or psychological energy.

An example of an interchange of energy occurs when an employee is trapped by a moving part of a machine, e.g. a power press. If the press is inadequately guarded, and the employee is able to get his hand into the danger area, injury results as the energy interchange is above the tolerance level of his hand.

(b) An organisation's financial well-being being placed at risk because of deficiencies in management; deficiencies in design and/or production capabilities; deficiencies in product quality/conformance; inability to expand and/or change; lack of adequate human resources; poor financial stability; lack of market appreciation/penetration; lack of awareness of cultural/social responsibilities; and failure to meet legal obligations.

An example of a risk to an organisation's well-being is the serving by an enforcement officer (e.g. HSE Inspector) of a prohibition notice when a risk of imminent danger exists. The prohibition notice stops the risky process, machine, department, or – in some cases – the whole business. As it is not possible to insure against the consequential financial loss arising from such a stop notice, the loss must be financed from within the organisation's profitability. If profit margins are tight, then an overall loss situation may result, thus threatening the overall financial viability of the organisation as a whole.

1.3 Risk management

Risk management may be defined as the eradication or minimisation of the adverse effects of the pure risks to which an organisation is exposed.

Pure risks can only result in a loss to the organisation, whereas with speculative risks, either gain or loss may result.

An example of a pure – or static – risk concerns a build-up of combustible material in the corner of a large distribution warehouse. If a source of ignition is present in the vicinity, then the risk of fire spread is greatly enhanced by the build-up of combustible material, thus posing the threat of a large loss of stock caused by fire. There will also be consequential loss resulting from the fire to consider, e.g. loss of profit on goods in stock; loss of market share etc.

An example of a speculative – or dynamic – risk concerns commodity purchasing. A company – speculating to accumulate – buys in quantities of a key raw material at price £x/tonne, as the price is favourable, hoping that a cost saving will be made, as the price is likely to increase in the future. The speculative risk may result in gain or loss, as the price of the raw material could continue to fall, after the purchase price has been agreed with the supplier. Alternatively, the price may rise to £(x + y)/tonne, thus achieving the anticipated saving.

It should be borne in mind that the division between 'pure' and 'speculative' risks is not absolute. For example, the speculative risk of operating a machine without an adequate guard has a number of pure risks associated with it – injury; death; prohibition notices; prosecution; fines; loss of profit etc.

The principles of a risk management programme are: risk identification, risk evaluation and risk control. These three principles have been enshrined in recent health and safety legislation – e.g. Lead, Asbestos, COSHH, Noise, Manual Handling and Display Screen Equipment – and have brought risk management strategies and legislative compliance together.

Indeed, the Management of Health and Safety at Work Regulations 1992[4] have now ensured that risk identification, evaluation – i.e. assessment – and control become the cornerstone of all health and safety management systems.

1.3.1 Risk management – role and process

The role[5] of risk management in industry and commerce is to:

1 consider the impact of certain risky events on the performance of the organisation;
2 devise alternative strategies for controlling these risks and/or their impact on the organisation; and
3 relate these alternative strategies to the general decision framework used by the organisation.

The process of risk management involves: identification, evaluation and control.

Risk identification may be achieved by a multiplicity of techniques, including physical inspections, management and worker discussions, safety audits, job safety analysis, and Hazop studies. The study of past accidents can also identify areas of high risk.

Risk evaluation (or measurement) may be based on economic, social or legal considerations.

Economic considerations should include the financial impact on the organisation of the uninsured cost of accidents, the effect on insurance premiums, and the overall effect on the profitability of the organisation and the possible loss of production following the issue of Improvement and Prohibition Notices.

Social and humanitarian considerations should include the general well-being of employees, the interaction with the general public who either live near the organisation's premises or come into contact with the organisation's operations – e.g. transportation, nuisance noise, effluent discharges etc. – and the consumers of the organisation's products or services, who ultimately keep the organisation in business.

Legal considerations should include possible constraints from compliance with health and safety legislation, codes of practice, guidance notes and accepted standards, plus other relevant legislation concerning fire prevention, pollution, and product liability.

The probability and frequency of each occurrence, and the severity of the outcome – including an estimation of the maximum potential loss – will also need to be incorporated into any meaningful evaluation.

1.3.2 Risk control strategies

Risk control strategies may be classified into four main areas: risk avoidance, risk retention, risk transfer and risk reduction.

1 Risk avoidance

This strategy involves a conscious decision on the part of the organisation to avoid completely a particular risk by discontinuing the operation producing the risk and it presupposes that the risk has been identified and evaluated.

For example, a decision may be made, subject to employees' agreement, to pay all wages by cheque or credit transfer, thus obviating the need to have large amounts of cash on the premises and the inherent risk of a wages snatch.

Another example of a risk avoidance strategy – from the health and safety field – would be the decision to replace a hazardous chemical by one with less or no risk potential.

2 Risk retention

The risk is retained in the organisation where any consequent loss is financed by the company. There are two aspects to consider under this heading: risk retention with knowledge, and risk retention without knowledge.

(a) With knowledge

This covers the case where a conscious decision is made to meet any resulting loss from within the organisation's financial resources. Decisions on which risks to retain can only be made once all the risks have been identified and effectively evaluated.

(b) Without knowledge

Risk retention without knowledge usually results from lack of knowledge of the existence of a risk or an omission to insure against it, and this often arises because the risks have not been either identified or fully evaluated.

3 Risk transfer

Risk transfer refers to the legal assignment of the costs of certain potential losses from one party to another. The most common way of effecting such transfer is by insurance. Under an insurance policy, the insurer (insurance company) undertakes to compensate the insured (organisation) against losses resulting from the occurrence of an event specified in the insurance policy (e.g. fire, accident etc.).

The introduction of clauses into sales agreements whereby another party accepts responsibility for the costs of a particular loss is an alternative risk transfer strategy. However, it should be noted that the conditions of the agreement may be affected by the Unfair Contract Terms Act 1977 and the interpretation placed on 'reasonableness'.

4 Risk reduction

The principles of risk reduction rely on the reduction of risk within the organisation by the implementation of a loss control programme, whose basic aim is to protect the company's assets from wastage caused by accidental loss.

The collection of data on as many loss producing accidents as possible provides information on which an effective programme of remedial action can be based. This process will involve the investigation, reporting and recording of accidents that result in either injury or disease to an individual, damage to property, plant, equipment, materials, or the product; or those near-misses where although there has been no injury, disease or damage, the risk potential was high.

The second stage of the development towards risk reduction is achieved by bringing together all areas where losses arise from accidents – whether fire, security, pollution, product liability, business interruption etc. – and co-ordinating action with the aim of reducing the loss. This risk reduction strategy is synonymous with loss control.

Loss control (or risk reduction) may be defined as a management system designed to reduce or eliminate all aspects of accidental loss that lead to a wastage of an organisation's assets.

As the emphasis on the economic argument increased, the technique of loss control has become more closely allied to financial matters, and in particular insurance.

The bringing together of insurance (risk transfer) and loss control (risk reduction) was the final stage in the development of the new discipline of risk management.

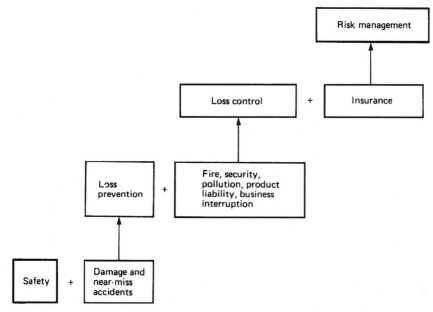

Figure 1.1 Pictogram showing development from safety to risk management.

This logical, progressive development from safety to risk management may be presented in pictogram form as in *Figure 1.1*.

1.4 Loss control

1.4.1 Introduction

Loss control may be defined as a management system designed to reduce or eliminate all aspects of accidental loss that may lead to a wastage of the organisation's assets. Those assets include manpower, materials, machinery, methods, manufactured goods and money. Loss control is based mainly on the economic approach to accident prevention, and loss control management is essentially the application of sound management techniques to the identification, evaluation and economic control of losses within a business. It has been shown in section 1.3.2 that loss control is synonymous with risk reduction, so the practical techniques associated with each stage of the process – i.e. identification, evaluation and control – are closely related.

Bird and Loftus[6] state that loss control management involves the following:

1 The identification of risk exposure.
2 The measurement and analysis of exposures.
3 The determination of exposures that will respond to treatment by existing or available loss control techniques or activities.

4 The selection of appropriate loss control action based on effectiveness and economic feasibility.
5 The managing of the loss control programme implementation in the most effective manner subject to economic constraints.

1.4.2 Component parts of a loss control programme

The component parts of a loss control programme may be considered in terms of protecting one or more of the organisation's assets from accidental loss, and will generally include: injury prevention (safety); damage control; fire prevention; security; industrial health and hygiene; pollution; product liability; and business interruption.

Injury prevention is concerned directly with the protection of the manpower asset within an organisation. To a lesser extent, it is indirectly concerned with the protection of the money asset, as a reduction in the number of injuries should result in a reduction in both the insured and the uninsured accident costs. In certain injury prevention programmes, the protection of the manpower asset is extended outside the factory, via the inclusion of off-the-job safety, road safety and home safety.

The following specific areas need to be examined in an injury prevention programme: safety policies; safety training; safety audits; identification of hazards; accident reporting and investigation; safe systems of work; machine and area guarding; housekeeping; personal protective equipment; and legislative compliance.

Damage control is directly concerned with the protection of assets comprising machinery, materials and manufactured goods from accidental loss before they reach the customer. Indirectly, this leads to the protection of the money asset through the elimination of repetitive damage and associated repair/replacement costs. Also, there may be some indirect protection of the manpower asset if damage causes and injury causes are similar. Essentially, damage control is an extension of the injury reporting and prevention programme to encompass also those accidents which result in damage only to plant, property, equipment and materials. The practices and techniques of damage control are considered in depth by Bird and Germain[7], and Bird and Loftus[8].

Incident recall[9,10] is another technique that can be utilised in a damage control programme to gain information about near-miss accidents.

Essentially, the incident recall technique may be used to identify unsafe acts, unsafe conditions, non-compliance with safe systems of work, and near-miss accidents by following a *confidential* interviewing procedure to a stratified random sample of employees. Each interviewee is asked to recall and report verbally any of the above-mentioned situations in which he was involved or has knowledge. Details of near-miss accidents are then obtained to enable remedial action to be taken *before* further similar accidents result in both damage and injury.

Fire prevention may be considered to be a special aspect of damage control in that it protects the machinery, materials and manufactured goods assets. It also protects the manpower asset, since fire can cause injury as well as damage and because fire damage is a very costly item, it is indirectly protecting the money asset as well. The Association of British Insurers has stated that the total cost fire insurance claims in the UK for 1997 was £1.1 bn, up 3% on 1996.

From a practical viewpoint, consideration should be given to aspects of fire prevention techniques, methods of fire control, firefighting and extinguishment, fire protection including fixed equipment (e.g. sprinklers etc.), storage and handling of flammable liquids, fire safety of employees, means of escape, evacuation drills and procedures, explosion potential, handling, storage and use of explosives, electrical installations, disaster contingency planning and the requirements of the Fire Precautions Act 1971 and the Highly Flammable Liquids and Liquefied Petroleum Gases Regulations 1972 and, more recently, the Fire Precautions (Workplace) Regulations 1997.

Security protects the materials, methods, manufactured goods and money assets. Its inclusion in a loss control programme is primarily based on economic considerations, as any breaches of security that result in losses of the organisation's assets may not be considered by the organisation to be accidental in nature.

However, certain safety hazards arise because of lack of security: for example, the high risk involved when employees are sent to collect wages; or the potential risk to children and others when a factory or construction site is not physically secure in terms of preventing unauthorised access.

Hence, indirectly, a system of security can improve the overall safety of the organisation, as well as directly protecting the assets mentioned above. Security – in the form of locked doors – can, however, sometimes conflict with the protection of the manpower asset, especially if the locked doors are the emergency fire exits.

The loss control programme should examine such areas as the physical security of premises; cash collection, handling, and distribution; theft and pilfering; vandalism; storage of valuable and attractive items; sabotage; industrial espionage; and the control of confidential data and methods. Consideration should also be given to defensive techniques, such as stocktaking, accounting/auditing checks, and the use of mechanical/ electrical safeguards. Aspects of computer security should also be included in this area, especially in the light of the increasing number of cases being reported that have involved some form of fraud, embezzle- ment or espionage utilising the computer.

Occupational health and hygiene is concerned with the protection of the manpower asset from the effects of occupational diseases – i.e. long-term accidents and other adverse conditions associated with the industrial environment. Indirectly, this also protects the money asset, as an improvement in health and hygiene within a factory should lead to a reduction in the incidence of occupational diseases, and hence a reduction in the associated costs.

Specific areas for consideration include: noise, dusts, gases, vapours, corrosives, toxic materials, radioactive materials, ventilation, heating, lighting, humidity, environmental monitoring, biological monitoring, health checks, general and personal hygiene, health education, counsel- ling, and employment/pre-employment medicals. Compliance with the

COSHH Regulations is vitally important to ensure a healthy workforce as is consideration of physical and mental stressors, as well as the more obvious chemical hazards.

Pollution control/environmental protection in all its aspects is concerned not only with the environment within the factory, but also the environment outside and around the factory. Control of air, ground and water pollution protects the manpower asset directly and the money asset indirectly. Adverse publicity resulting from an organisation causing some form of pollution would initially harm the organisation's image, and perhaps harm it economically. Persistent breaches of one or other of the Acts dealing with pollution can ultimately lead to prosecution and fines with further adverse publicity, both at local and national level.

Special attention needs to be paid to the Control of Pollution Act 1974 that deals with the control of noise as a pollutant, in addition to the other areas of air, ground and water pollution. All possible areas and types of pollution should be identified as part of the loss control audit. With the advent of the Environmental Protection Act 1990[11] and BS 7750:1992: Environmental Management Systems[12] – now BS EN ISO 14001: 1996[20] – the environmental aspects of the loss control programme have increasingly gained in importance in recent years. Indeed, many health and safety practitioners now have the added responsibility to provide advice on environmental matters.

Product liability extends the protection to all consumers of the organisation's products or services, and is therefore primarily concerned with the protection of the money asset. This asset may suffer accidental losses directly because of increased insurance premiums that reflect large compensation payments, or indirectly because of adverse publicity which is detrimental to the organisation's image.

Practical considerations in this area should centre on the development of a product safety strategy that is in keeping with both occupational (section 6, Health and Safety at Work etc. Act 1974) and consumer (Consumer Protection Act 1987) product safety legislation, codes of practice and guidance notes.

Areas to be considered should include: products loss control policy; products loss control committee chaired by senior manager and comprising designers, R and D personnel, manufacturing and production management, safety advisers, quality control, servicing, sales and advertising, and distribution; product safety incorporated at the design and R and D stages; written quality control/assurance procedures, i.e. BS 5750: 1987[13] – now BS EN ISO 9001 1994[21]; role of sales, marketing, advertising, distribution and servicing personnel in product safety; complaints system; and products recall system.

Business interruption or continuity further extends the loss control strategy to take account of the fact that time is money, and, as such, any loss of production or service is detrimental to the overall profitability of the company. Hence, business interruption is primarily concerned with the protection of the money asset. Indirectly, however, it serves to maintain the assets of machinery, materials, manufactured goods and methods.

A programme to prevent business interruption can include planned lubrication, planned preventive maintenance, condition monitoring, statutory inspections, machinery replacement programmes, availability of key spares, identification of key machines, processes, areas, personnel etc. within the organisation, continued supply of raw materials, minimisation of production bottlenecks, and highlighting dependencies on specific items of plant, suppliers, customers, personnel, computer systems, and/or public utilities (e.g. gas, electricity).

1.4.3 Loss control management in practice

The aforementioned areas – from injury prevention to business interruption – require to be co-ordinated within one senior management function (possibly risk management) in order to ensure a rational and concerted approach to the problem of eliminating or reducing the costly accidental losses that can occur within an organisation. These so-called operational losses inevitably lead to an erosion of profit margins, and also adversely affect the overall performance of the organisation.

This co-ordinated management role is crucial to the success of any loss control programme. The senior manager responsible for programme implementation should have authority to make decisions and take action without the need to seek day-to-day approval for his decisions. He should report on a regular (monthly) basis to the main board on the implementation of the loss control programme within the organisation. Without the backing and commitment of the most senior executives, it is doubtful whether a programme can be successfully introduced.

An effective programme of loss control (risk reduction) not only leads to a more profitable situation, but will also greatly assist legislative compliance, and will result in a reduction in the total number of accidents within the organisation's operations.

1.5 Degrees of hazard

An awareness of the differing degrees of hazard to people will enable appropriate control measures to be developed and implemented.

Immediate physical danger can manifest itself through very short-term injury accidents – e.g. hand amputation in a power press; person falling from a height. The result of immediate physical danger – if it goes uncontrolled – will inevitably be *immediate physical injury*. The enforcing agencies use the phrase 'risk of imminent danger' or 'risk of serious personal injury' in connection with the issuing of prohibition notices – a legal control measure designed to reduce the risk of immediate physical danger.

Long-term physical danger is more cumulative or chronic than acute or short term. Cumulative back strain caused by poor kinetic handling techniques is an example of *long-term physical injury*.

Immediate chemical danger may be caused by strong acids and alkalis being poorly stored and handled, thus leading to a risk of skin contact and corrosive burns – i.e. *immediate chemical injury.*

Long-term chemical danger is again chronic or cumulative – e.g. lead poisoning or exposure to asbestos fibres. The result is some form of occupational disease – i.e. *long-term chemical injury.*

Immediate biological danger may be caused by the presence of contagious diseases or via genetic manipulation. The result is again some form of occupation disease or illness.

Long-term biological danger is usually cumulative in nature, for example noise-induced occupational deafness.

Immediate psychological danger is linked to short-term trauma – e.g. a disaster at home or work; social problems – domestic illness etc. This may result in a loss of concentration, abruptness with work colleagues, and other short-term stress-related symptoms.

Long-term psychological danger may be linked to fears connected with fear of failure, unemployment/job security, or lack of career direction and motivation. The symptoms are similar to those described above, but often only become apparent over a longer timescale.

1.6 Accident causation models

1.6.1 Sequence of events – domino theory

The 'Domino Theory' attributed to Heinrich[14] is based on the theory that a chain or sequence of events can be listed in chronological order to show the events leading up to an accident:

event a → event b → event c → accident → effect

Each event may have more than one cause, i.e. be multicausal.

Heinrich states that the occurrence of an injury accident invariably results from a completed sequence of factors culminating in the accident itself. He postulates five factors or stages in the accident sequence, with the injury invariably caused by the accident, and the accident in turn the result of the factor that immediately precedes it.

The five factors or stages in the sequence of events are:

(a) ancestry and social environment, leading to
(b) fault of person, constituting the proximate reason for
(c) an unsafe act and/or mechanical hazard, which results in
(d) the accident, which leads to
(e) the injury.

Heinrich likens these five stages to five dominoes standing on edge in a line next to each other, so that when the first domino falls it automatically knocks down its neighbour which in turn knocks down its neighbour and so on. Removal of any one of the first four will break the sequence and so prevent the injury.

In fact, Heinrich suggested that accident prevention should aim to remove or eliminate the middle or third domino, representing the unsafe act, mechanical or physical hazard, thus preventing the accident.

During accident investigations, in addition to asking 'What action has been taken to prevent recurrence?', the investigator needs to be aware of the chain of events leading up to the accident, and tracing it back. Similarly, on safety audits and inspections, when the risk of an accident has been identified, possible event chains should be investigated and action taken to remove potential causes.

1.6.2 An updated domino sequence

Bird and Loftus[15] have extended this theory to reflect the influence of management in the cause and effect of all accidents that result in a wastage of the company's assets. The modified sequence of events becomes:

(a) lack of control by management, permitting
(b) basic causes (personal and job factors), that lead to
(c) immediate causes (substandard practices/conditions/errors), which are the proximate causes of
(d) the accident, which results in
(e) the loss (minor, serious or catastrophic).

This modified sequence can be applied to all accidents, and is fundamental to loss control management.

1.6.3 Multiple causation theory

Multicausality refers to the fact that there may be more than one cause to any accident:

Each of these multicauses is equivalent to the third domino in the Heinrich theory and can represent an unsafe act or condition or situation. Each of these can itself have multicauses and the process during accident investigation of following each branch back to its root is known as 'fault tree analysis'.

The theory of multicausation is that the contributing causes combine together in a random fashion to result in an accident. During accident investigations, there is a need to identify as many of these causes as possible. In reality, the accident model is an amalgam of both the domino and multicausality theories.

Petersen has compared and contrasted both theories and gives an example[16] which illustrates the comparative narrowness of the domino theory in relation to the multicausality theory and concludes that this has severely limited the identification and control of the underlying causes of accidents.

The theory of multicausality has its basis in epidemiology. Gordon[17] points out that accidental injuries could be considered with epidemiological techniques. He believes that if the characteristics of the 'host' (accident victim), of the agent (the injury deliverer), and of the supporting 'environment' could be described in detail, more understanding of accident causes could be achieved than by following the domino technique of looking for a single cause only. Essentially, Gordon's theory is that the accident is the result of a complex and random interaction between the host, the agent and the environment, and cannot be explained by consideration of only one of the three.

1.6.4 Failure modes and effects

This technique involves a sequential analysis and evaluation of the kinds of failures that could happen, and their likely effects, expressed in terms of maximum potential loss.

The technique is used as a predictive model and would form part of an overall risk assessment study.

1.6.5 Fault tree analysis

Fault tree analysis is an analytical technique that is used to trace the chronological progression of factors (events) contributing to the accident situation, and is useful in accident investigation and as a predictive, quantitative model in risk assessment. Again, the principle of multicausality is utilised in this type of analysis. (A fuller treatment on fault tree analysis is given at section 10.6.)

1.7 Accident prevention: legal, humanitarian and economic reasons for action

1.7.1 Introduction

In order to get action taken in the field of accident prevention, safety advisers have the three fundamental lines of attack on which to base their strategies for generating and maintaining management activity in this area. These three reasons for accident prevention make use of the legal, humanitarian and economic arguments respectively. An optimum accident prevention strategy for a particular organisation would involve a combination of the three, because they are interrelated and probably reinforce one another.

1.7.2 Legal reasons for accident prevention

The legal argument is based on the statutory requirements of the HSW, FA and other related legislation.

The HSW imposes a general duty on employers to ensure, so far as is reasonably practicable, the health, safety and welfare of all his employees. The term 'reasonably practicable' involves balancing the cost of preventing the accident against the risk of the accident occurring. Thus, economic considerations need also to be taken into account.

PUWER lays down more specific statutory requirements which impose a minimum but absolute standard of conduct on the employer.

Any breach of the statutory duties imposed by either of the aforementioned Acts can result in the employer being involved in criminal proceedings. The penalties under the Health and Safety at Work Act include unlimited fines and imprisonment for up to two years, for prosecution on indictment. On average, 20 directors, managers, supervisors, employees are individually prosecuted per annum. A number of individuals have been given custodial sentences under, for example, asbestos, machinery safety and gas safety legislation. The maximum fine on summary conviction for certain offences is currently (May 1998) £20 000 with the maximum for other offences being £5 000.

The safety adviser can therefore reason via the legal argument for accident prevention on the basis that the employer should avoid attracting a prosecution.

The economic argument is also relevant here, because of the fines that may be imposed as a result of statutory breaches, and also because of the impact of Improvement and Prohibition Notices in terms of uninsurable consequential loss arising out of enforced cessation of work.

The image of the company or organisation is also likely to be tarnished as a result of adverse publicity received in connection with any prosecution for breaches of statute. Loss of company image has predominantly economic disadvantages, usually because of the loss of good will or other intangible and invisible company assets, which in turn indirectly leads to a loss of business.

1.7.3 Humanitarian reasons for accident prevention

The humanitarian reason for accident prevention is based on the notion that it is the duty of any man to ensure the general well-being of his fellow men. This places an onus on the employer – the common law duty of care – to provide a safe and healthy working environment for all his employees.

An illustration of this occurs in the case of *Wilsons and Clyde Coal Co. Ltd v. English*[18], where Lord Wright said that 'the whole course of authority consistently recognises a duty which rests on the employer, and which is personal to the employer, to take reasonable care for the safety of his workmen, whether the employer be an individual, a firm or a company and whether or not the employer takes any share in the conduct of the operations'.

There is some overlap here between common and statute law, as the Health and Safety at Work Act places a general duty on an employer to ensure, so far as is reasonably practicable, the health, safety and welfare of his employees.

The safety adviser is therefore able to argue – via humanitarian reasoning – that it is immoral to have a process or machine which may injure employees, and he can stress the possible outcome of such dangers in terms of pain and suffering.

1.7.4 Economic reasons for accident prevention

The fundamental reason for utilising the economic argument in the promotion of accident prevention is the fact that accidents cost an organisation money. However, in order to press the economic argument, knowledge is needed of the costs to the organisation of all types of accident.

Essentially, there are two types of accident costs – the insured costs, and the uninsured costs.

The insured (or direct) costs are predominantly covered by the Employer's Liability insurance premium, which to all intents and purposes is the direct accident cost to the majority of organisations.

The uninsured (or indirect, hidden) costs of accidents should also be established. Bamber[19] developed a list of uninsured costs which is considered to be objective, and which will readily be accepted by operational management as being costs associated with accidents:

1 Safety administration and accident investigation.
2 Medical and treatment.
3 Cost of lost time of injured person.
4 Cost of lost time of other employees.
5 Cost of replacement labour.
6 Cost of payments to injured person.
7 Cost of loss of production and business interruption.
8 Cost of repair to damaged plant.
9 Cost of replacement of damaged materials.
10 Other costs – e.g. photographs, transport, accommodation, wage details, fees etc.

The above list of costs should be utilised in the calculation of the total accident costs to the organisation, to enable senior management to gauge the relative impact of such costs, by comparing them with other business costs.

The safety adviser is therefore able to reason – via the economic argument – that accident prevention may well be cost-effective. But the organisation is reducing pain and suffering by having an effective system of accident prevention, as well as saving money. Thus, the economic argument gives support to both the legal – via the use of economic sanctions – and the humanitarian arguments. In order to achieve maximum co-operation in any programme of accident prevention, use should be made of an amalgam of all three arguments, i.e. legal, humanitarian and economic. However, from a motivational point of view, it is the economic argument that has the greatest impact with directors and senior management.

References

1. Ministry of Labour and National Service, *Industrial Accident Prevention*, Report of the Industrial Safety Sub-Committee of the National Joint Advisory Council (1956)
2. Health and Safety Executive, *Hazard and Risk Explained – Control of Substances Hazardous to Health Regulations 1988* (COSHH), Leaflet No. IND(G)67(L), HSE Books, Sudbury (1988)
3. Health and Safety Commission, Legal Series booklet No. L21, *Management of Health and Safety at Work Regulations 1992: Approved Code of Practice*, 3, HSE Books, Sudbury (1992)
4. *Management of Health and Safety at Work Regulations 1992*, HMSO, London (1992)
5. Carter, R. L. and Doherty, N., *Handbook of Risk Management*, 1.1–06, Kluwer-Harrap, London (1974)
6. Bird, F. E. and Loftus, R. G., *Loss Control Management*, 52, Institute Press, Longanville, Georgia (1976)
7. Bird, F. E. and Germain, G. L., *Damage Control*, American Management Association, New York (1966)
8. Ref. 6, pp. 93–138
9. Ref. 6, pp. 215–246
10. Bamber, L., Incident recall – a (lack of) progress report, *Health and Safety at Work*, **2**, No. 9, 83 (1980)
11. *The Environmental Protection Act 1990*, HMSO, London (1990)
12. British Standards Institution, *BS 7750:1992 Specification for Environmental Management Systems*, BSI, London (1992)
13. BS 5750: Parts 1–6:1987, *Quality systems*, British Standards Institution, London
14. Heinrich, H. W., *Industrial Accident Prevention*, 4th edn, 13–16, McGraw-Hill, New York (1959)
15. Ref. 6, pp. 39–48
16. Petersen, D. C., *Techniques of Safety Management*, 2nd edn, 16–19, McGraw-Hill, Kogakusha, USA (1978)
17. Gordon, J. E., The epidemiology of accidents, *Amer. J. of Public Health*, **39**, 504–515 (1949)
18. Wilsons and Clyde Coal Co. Ltd *v.* English, (1938) AC 57 (HL)
19. Bamber, L., Accident prevention the economic argument, *Occupational Safety and Health*, **9**, No. 6, 18–21 (1979)
20. British Standards Institution BS EN ISO 14001: 1996, *Environmental management systems – Specification with guidance for use*, BSI, London (1996)
21. British Standards Institution BS EN ISO 9001: 1994, *Quality systems – Specification for design/development, production, installation and servicing*, BSI, London (1994)

Further reading

Heinrich, H.W., Petersen, D. and Roos, N., *Industrial Accident Prevention – A Safety Management Approach*, 5th edn, McGraw-Hill, New York (1980)

DeReamer, R., *Modern Safety Practices*, John Wiley & Sons Inc., New York (1958)

Bird, F.E. and Loftus, R.G., *Loss Control Management*, Institute Press, Loganville, Georgia (1976)

Petersen, D.C., *Techniques of Safety Management*, 2nd edn, McGraw-Hill, Kogakusha, USA (1978)

Hale, A.R. and Hale, M., *A Review of the Industrial Accident Research Literature*, Committee on Safety and Health at Work: Research Paper, HMSO, London (1972)

Crockford, G.N., *An Introduction to Risk Management*, Woodhead-Faulkner, Cambridge (1980)

Carter, R.L. *et al.*, *Handbook of Risk Management*, Kluwer Publishing Ltd, Kingston-upon-Thames (1997–1998)

Health and Safety Executive Publication No. HS(G)65 *Successful Health and Safety Management* (2nd ed), HSE Books, Sudbury (1997)

British Standards Institution BS 8800: 1996, *Guide to Occupational health and safety management systems*, BSI, London (1996)

Chapter 2

Risk management: techniques and practices

L. Bamber

2.1 Risk identification, assessment and control

2.1.1 Introduction

As discussed in section 1.2.2, the risk from a hazard is the likelihood that it will cause harm in the actual circumstances in which it exists.

Essentially, the technique of risk management involves:

1 identification
2 assessment
3 control (elimination or reduction).

Within the workplace, operational management at all levels has a responsibility to identify, evaluate and control risks that are likely to result in injury, damage or loss. Part of these responsibilities should involve implementation of a regular programme of safety inspections of the work areas under their control. These inspections should include physical examinations of the workplace – i.e. the nuts and bolts – and also the systems, procedures, and work methods – i.e. the organisational aspects.

The process of risk management has been briefly outlined in section 1.3.1. The following sections (2.1.2–2.1.4) consider the practical application of the techniques in the workplace.

2.1.2 Risk identification

Within an organisation, there are several ways by which risks may be identified. These include:

1 Workplace inspections.
2 Management/worker discussions.
3 Independent audits.
4 Job safety analysis.
5 Hazard and operability studies.
6 Accident statistics.

Workplace inspections are undertaken with the aim of identifying risks and promoting remedial action. Many different individuals and groups within an organisation will – at some time – be involved in a workplace inspection: directors, line managers, safety adviser, supervisors and safety representatives. The key aspect is that results of all such inspections should be co-ordinated by one person within the factory, whose responsibility should include (a) monitoring action taken once the risk has been notified, and (b) informing those persons who reported the risk as to what action has been taken.

The vast majority of workplace inspections concentrate on the 'safe place' approach – i.e. the identification of unsafe conditions – to the detriment of the 'safe person' approach – i.e. the identification of unsafe acts.

Heinrich states that only 10% of accidents are caused by unsafe mechanical and physical conditions, whereas 88% of accidents are caused by unsafe acts of persons. (The other 2% are classed as unpreventable, or acts of God!).

Hence for workplace inspections to be beneficial in terms of risk identification and accident prevention, emphasis *must* be placed on the positive safe person approach, using techniques such as:

- safe visiting – talking to people
- catching people doing something right (not wrong)
- positive behavioural reinforcement
- one-to-one training/counselling sessions,

as well as the more traditional safe place approach which tends to be more negative as it evokes fault finding and blame apportionment at all levels within an organisation – i.e. catching people doing something wrong and penalising them for it.

Workplace inspections tend to follow the same format but are given many different names including: safety sampling, safety audits, safety inspections, hazard surveys, etc. Certain of the above are discussed below but all have the same aim – namely risk identification.

Management/worker discussions can also be useful in the identification of risks. Formal discussions take place during meetings of the safety committee with informal discussions occurring during on-the-job contact or in conversations between supervisor and worker. The concept of incident recall[1,2] is an example of management/worker discussion.

Indeed, incident recall has in effect been given legal status via Regulation 12 of the Management of Health and Safety at Work Regulations 1992 which requires employees to highlight shortcomings in systems and procedures – i.e. hazards, defects, damage and near-miss accidents, unsafe conditions and unsafe activities.

In all cases, however, the feedback element is important from a motivational viewpoint. The risk identifier must be kept fully informed of any action taken to prevent injury, damage or loss arising from the risk he has noted.

Independent audits can also be used to identify risks. The term 'independent' here refers to those who are not employees of the organisation, but who – from time to time – undertake either general or specific workplace audits or inspections. Such independent persons may include:

1 Engineer surveyors – insurance company personnel undertaking statutory inspections of boilers, pressure vessels, lifting tackle etc. They are employed by the organisation as 'competent persons'.
2 Employers' liability surveyors – insurance company personnel undertaking general health and safety inspections in connection with employers' liability insurance.
3 Claims investigators – insurance company personnel investigating either accidents in connection with injury or damage claims under insurance policies.
4 Insurance brokers personnel – risk management or technical consultants undertaking inspections in connection with health and safety, fire, or engineering insurance as part of client servicing.
5 Outside consultants – undertaking specific investigations on a fee-paying basis. For example, noise or environmental surveys may be commissioned, if the expertise is not available within the organisation. Trade associations may be of assistance in this area.
6 Health and Safety Executive – factory (and other) inspectors undertaking either general surveys or specific accident investigations.

Again, with all the above there is a need to co-ordinate their independent findings to ensure that action is promptly taken to control any risks identified.

Job safety analysis is another method of risk identification. A fuller discussion of this method is presented below (see section 2.2.1).

Hazard and operability studies are useful as a risk identification technique, especially in connection with new designs/processes. The technique was developed in the chemical process industries, and essentially it is a structured, multi-disciplinary brainstorming session involving chemists, engineers, production management, safety advisers, designers etc. critically examining each stage of the design/process by asking a series of 'what if?' questions. The prime aim is to design out risk at the early stages of a new project, rather than have to enter into costly modifications once the process is up and running.

Further information on Hazop studies may be found in the Chemical Industries Association's publication on the subject[3].

Accident statistics will be useful in identifying uncontrolled risks as they will present – if properly analysed from a causal viewpoint – data indicative of where control action should have been taken to prevent recurrence. Ideally, an analysis of *all* injury, damage and near-miss accidents should be undertaken, so that underlying trends may be highlighted and effective control action – both organisational and physical in nature – taken.

2.1.3 Risk assessment

Once a list of risks within a company has been compiled, the impact of each risk on the organisation – assuming no control action has been taken

– requires assessment, so that the risks may be put in order of priority in terms of when control action is actually required, i.e. immediate; short term; medium term; long term on the basis of a ranking of the risks relating to their relative impact on the organisation. Such an assessment should take account of legal, humanitarian and economic considerations (as outlined in section 9.7).

The fundamental equation in any risk assessment exercise is:

Risk magnitude = Frequency (how often?) × Consequence (how big?)

In general:

- Low-frequency, low-consequence risks should be retained (i.e. self-financed) within the organisation. Examples include the failure of small electric motors, plate-glass breakages, and possibly motor vehicle damage accidents (via retention of comprehensive aspects of insurance cover).
- Low-frequency, high-consequence risks should be transferred (usually via insurance contracts). Examples include explosions, and environmental impairment.
- High-frequency, low-consequence risks should be reduced via effective loss control management. Examples include minor injury accidents; pilfering; and damage accidents.
- High-frequency, high-consequence risks should (ideally) be avoided by managing them out of the organisation's risks portfolio. If this appears to be an uneconomic (or unpalatable) solution, then adequate insurance – i.e. the risk transfer option – *must* be arranged.

A quantitative method of risk assessment – which takes into account the risk magnitude equation discussed above – considers the frequency (number of times spotted); the maximum potential loss (MPL) – i.e. the severity of the worst possible outcome; and the probability that the risk will actually come to fruition and result in a loss to the organisation.

From this type of quantitative assessment, a list of priorities for risk control can be established, and used as a basis to allocate resources.

A simple risk assessment formula involving frequency, MPL and probability is:

Risk rating = Frequency × (MPL + probability)

In the above formula, frequency (F) is the number of times that a risk has been identified during a safety inspection.

Maximum Potential Loss (MPL) is rated on a 50-point scale where, for example:

multifatality	– 50
single fatality	– 45
total disablement	– 40
(para/quadraplegic)	

loss of eye	– 35
arm/leg amputation	– 30
hand/foot amputation	– 25
loss of hearing	– 20
broken/fractured limb	– 15
deep laceration	– 10
bruising	– 5
scratch	– 1

Probability (P) is rated on a 50-point scale where, for example:

imminent	– 50
hourly	– 35
daily	– 25
once per week	– 15
once per month	– 10
once per year	– 5
once per five or more years	– 1

Consider an example where the risk to be assessed has been identified once during an inspection. The MPL (worst possible outcome) was considered to be the loss of an eye with the probability of occurrence of once per day.

Thus, for this risk, the rate is:

$$RR = F \times (MPL + P)$$
$$= 1 \times (35 + 25)$$
$$= 60$$

This risk rating figure should then be compared to a previously agreed risk control action guide, such as:

Risk rating	Urgency of action
Over 100	Immediate
80–100	Today
60–79	Within 2 days
40–59	Within 4 days
20–39	Within 1 week
10–19	Within 1 month
0–9	Within 3 months

These action scales should be drawn up by individual organisations, taking into account both the human and financial resources available for risk control.

In our example, the risk rating was found to be 60, hence control action to eliminate (or reduce) the risk should be taken within two days.

The above scales, example and action guide serve only to illustrate the principles involved, and – because of resource constraints – may not be generally applicable for practical use in all organisations.

2.1.4 Risk control

The four risk control strategies – avoidance, retention, transfer and reduction – have been discussed in section 9.3.2.

The bulk of the risks identified by regular safety inspections will require some form of risk reduction (or avoidance) through effective loss control management.

The control of risks within an organisation requires careful planning, and its achievement will involve both short-term (temporary) and long-term (permanent) measures.

These measures can be graded thus:

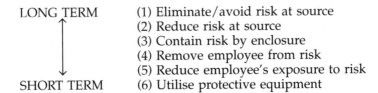

LONG TERM   (1) Eliminate/avoid risk at source
                           (2) Reduce risk at source
                           (3) Contain risk by enclosure
                           (4) Remove employee from risk
                           (5) Reduce employee's exposure to risk
SHORT TERM   (6) Utilise protective equipment

The long-term aim must always be to eliminate the hazard at source, but, whilst attempting to achieve this aim, other short-term actions – for example, utilisation of protective equipment – will be necessary. This list indicates an 'order-of' priority for remedial measures for any risk situation.

Various techniques are available to control risks within the workplace.

Mechanical risks may be engineered out of the process, or effectively enclosed by means of fixed guarding. Alternative forms of guarding involve the use of interlocked guards, light-sensitive barriers or pressure-sensitive mats. Trip devices and other forms of emergency stops may also be incorporated.

Risks from the working environment may be controlled by effective ventilation systems, adequate heating and lighting, and the general provision of good working conditions.

Chemical risks may also be controlled by effective ventilation, regular monitoring, substitution of material, change of process, purchasing controls, and the use of protective equipment.

A necessary corollary of risk assessment is the establishment of safe systems of work and training for the workforce to make them aware of the risks in their work areas, and of the methods for the control of such risks.

2.2 Job safety analysis

2.2.1 Job safety analysis – procedure

Job safety analysis (or job hazard analysis) is an accident prevention technique that should be used in conjunction with the development of job safety instructions; safe systems of work; and job safety training.

The technique of job safety analysis (JSA) has evolved from the work study techniques known as method study and work measurement.

The method study engineers' aim is to improve methods of production. In this they use a technique known as the SREDIM principle:

Select (work to be studied);
Record (how work is done);
Examine (the total situation);
Develop (best method for doing work);
Install (this method into the company's operations);
Maintain (this defined and measured method).

Work measurement is utilised to break the job down into its component parts and, by measuring the quantity of work in each of the component parts, make human effort more effective. From experience standard times have evolved for particular component operations and these enable jobs to be given a 'time'.

Job safety analysis uses the SREDIM principle but measures the risk (rather than the work content) in each of the component parts of the job under review. From this detailed examination a safe method for carrying out each stage of the job can be developed.

The basic procedure for job safety analysis is as follows:

1 Select the job to be analysed. (SELECT)
2 Break the job down into its component parts in an orderly and chronological sequence of job steps. (RECORD)
3 Critically observe and examine each component part of the job to determine the risk of accident. (EXAMINE)
4 Develop control measures to eliminate or reduce the risk of accident. (DEVELOP)
5 Formulate written and safe systems of work and job safety instructions for the job. (INSTALL)
6 Review safe systems of work and job safe practices at regular intervals to ensure their utilisation. (MAINTAIN)

From a practical viewpoint, this information can be recorded on a job safety analysis chart of the sort shown in *Figure 2.1*.

This is a typical job safety analysis chart. The detailed format will depend on the process and company and should be adapted to suit.

Criteria to be considered when selecting jobs for analysis will include:

1 past accident and loss experience;
2 maximum potential loss;
3 probability of recurrence;
4 legal requirements;
5 the newness of the job; and
6 the number of employees at risk.

Job Safety Analysis Record Chart		
Job title:	Date of job analysis:	
Department:	Time of job observation:	
Analyst/reviewer:		
Description of job:		
Accident experience:		
Maximum potential loss:		
Legal requirements:		
Relevant codes of practice/guidance notes/advisory publications:		
Sequence of job steps	Risks identified	Precautions advised
Suggested safe system of work:		
Suggested review date:		
Suggested job safety instructions:		
Suggested training programme:		
Signed:	Date:	
Department:	Function:	

Figure 2.1 Job safety analysis record chart

The ultimate aim must be to undertake JSA on *all* jobs within an organisation.

Once the job has been selected, the next stage is to break it down into its component parts or job steps. On average, there will be approximately 15 job steps; if more than 20, then the job under study should be sub-divided; if less than 10, then a bigger slice of the job should be analysed. Each job step should be one component part of the total job where something happens to advance by a measurable amount the doing of the work involved. The breakdown should be neither too general nor too specific. An example of such a job breakdown is given below.

Changing a car wheel

Job step	Risk factor	Control action
1 Put on handbrake	Strain to wrist/arm	Avoid snatching, rapid movement
2 Remove spare from boot and check tyre pressure	Strain to back	Use kinetic handling techniques
3 Remove hub cap	Strain; abrasion to hand	Ensure correct lever used
4 Ensure jack is suitable and is located on firm ground	Vehicle slipping. Jack sinking into ground	Check jack
5 Ensure jacking point is sound	Vehicle collapse	Consider secondary means of support
6 Jack up car part-way, but not so that the wheels leave the ground	Strain; bumping hands on jack/car	Avoid snatching, rapid movements
7 Loosen wheel-nuts	Hands slipping – bruised knuckles. Strain	Ensure spanner brace in good order. Avoid snatching, rapid movements. Use gloves
8 Jack up car fully in accordance with manufacturer's advice	Strain; bumping hands on jack/car	Avoid snatching, rapid movements
9 Remove wheel	Strain to back. Dropping onto feet	Use kinetic handling techniques. Use gloves (if available) to improve grip
10 Fit spare	Strain to back	Use kinetic handling techniques
11 Tighten wheel-nuts	Hand slipping – bruised knuckles. Strain	Use gloves. Avoid snatching, rapid movements

Changing a car wheel *(continued)*

Job step	Risk factor	Control action
12 Lower car	Strain; bumping hands on jack/car	Avoid snatching, rapid movements
13 Remove jack and store in boot, together with removed wheel	Strain to back	Use kinetic handling techniques
14 Retighten wheel-nuts	Hand slipping – bruised knuckles	Use gloves. Avoid snatching, rapid movements
15 Replace hub cap	Abrasion to hand	Use gloves
16 Ensure wheel is secure, prior to driving off		Check wheel and area around car

From the above, it may be seen that each job step has been systematically analysed for its component risk factor. For each identified risk factor a control action has been developed.

The third column – Control action – becomes the Job Safety Instructions, and forms the basis of the written safe system of work.

2.2.2 Job safety instructions

Once the individual job has been analysed, as described above, a written safe system of work should be produced.

The purpose of job safety instructions is to communicate the safe system of work to employees. For each job step, there is a corresponding control action designed to reduce or eliminate the risk factor associated with the job step. This becomes the job safety instruction which spells out the safe (and efficient) method of undertaking that specific job step.

Such job safety instructions should be utilised in as much job safety training both formal (in the classroom) and informal (on the job contact sessions) as possible. All managers and supervisors concerned should be fully knowledgeable and aware of the job safety instructions and safe systems of work that are relevant to the areas under their control.

From a practical viewpoint, job safety instructions should be listed on cards which should be (a) posted in the area in which the job is to be undertaken; (b) issued on an individual basis to all relevant employees; and (c) referred to and explained in all related training sessions.

2.2.3 Safe systems of work

Safe systems of work are fundamental to accident prevention and should:
(a) fully document the hazards, precautions and safe working methods,
(b) include job training, and (c) be referred to in the 'Arrangements'
section (part 3) of the Safety Policy.

Where safe systems of work are used, consideration should be given in
their preparation and implementation to the following:

1 Safe design.
2 Safe installation.
3 Safe premises and plant.
4 Safe tools and equipment.
5 Correct use of plant, tools and equipment (via training and
   supervision).
6 Effective planned maintenance of plant and equipment.
7 Proper working environment ensuring adequate lighting, heating and
   ventilation.
8 Trained and competent employees.
9 Adequate and competent supervision.
10 Enforcement of safety policy and rules.
11 Additional protection for vulnerable employees.
12 Formalised issue and proper utilisation of all necessary protective
   equipment and clothing.
13 Continued emphasis on adherence to the agreed safe method of work
   by all employees at *all* levels.
14 Regular (at least annually) reviews of all written systems of work to
   ensure:
   (a) compliance with current legislation,
   (b) systems are still workable in practice,
   (c) plant modifications are taken account of,
   (d) substituted materials are allowed for,
   (e) new work methods are incorporated into the system,
   (f) advances in technology are exploited,
   (g) proper precautions are taken in the light of accident experience,
      and
   (h) continued involvement in, and awareness of the importance of,
      written safe systems of work.
15 Regular feedback to all concerned – possibly by safety committees and
   job contact training sessions – following any changes in existing safe
   systems of work.

The above 15 points give a basic framework for developing and
maintaining safe systems of work.

2.3 System safety

2.3.1 Principles of system safety

A necessary prerequisite in connection with the study of system safety is
a working knowledge of the principles of safe systems of work and job

safety analysis. Also an appreciation of how hazard and operability studies[3] can be used will be of assistance.

System safety techniques have primarily emanated from the aviation and aerospace industries, where the overriding concern is for the complete system to work as it has been designed to, so that no one becomes injured as a result of malfunction.

Therefore, system safety techniques may be applied in order to eliminate any machinery malfunctions or mistakes in design that could have serious consequences. Thus, there is a need to analyse critically the complete system in order to anticipate risks, and estimate the maximum potential loss associated with such risks, should they not be effectively controlled.

The principles of system safety are founded on pre-planning and organisation of action designed to conserve all resources associated with the system under review.

According to Bird and Loftus[4], the stages associated with system safety are as follows:

1 The pre-accident identification of potential hazards.
2 The timely incorporation of effective safety-related design and operational specification, provisions, and criteria.
3 The early evaluation of design and procedures for compliance with applicable safety requirements and criteria.
4 The continued surveillance over all safety aspects throughout the total life-span – including disposal – of the system.

System safety may therefore be seen to be an ordered monitoring programme of the system from a safety viewpoint.

It may be seen that the system safety approach is very closely allied to the risk management approach. Indeed, the logical progression of system safety management techniques has been incorporated into many risk management processes, and also to other linked disciplines such as total quality management and environmental management systems.

2.3.2 The system

The system under review is the sum total of all component parts working together within a given environment to achieve a given purpose or mission within a given time over a given life-span.

The elements or component parts within a system will include manpower, materials, machinery and methods.

Each system will have a series of phases, which follow a chronological pattern; the sum total of which will equate to the overall life-span of the system. These phases are: conceptual phase, design and engineering phase, operational phase, and disposal phase:

1 The conceptual phase considers the basic purpose of the system and formulates the preliminary designs and methods of operation. It is at this stage that hazard and operability studies should be undertaken.

2 The design and engineering phase develops the basic idea from the conceptual phase, and augments them to enable translation into practical equipment and procedures. This phase should include testing and analysis of the various components to ensure compliance with various system specifications. It is at this stage that job safety analysis should be undertaken.

3 The operational phase involves the bringing together of the various components – i.e. manpower, materials, machinery, methods – in order to achieve the purpose of the system. From a practical viewpoint, it is at this stage that safe systems of work should be developed and communicated.

4 The disposal phase begins when machinery and manpower are no longer needed to achieve the purpose of the system. All components must be effectively disposed of, transferred, reallocated or placed into storage.

2.3.3 Method analysis

There are many methods of analysis in use in systems safety including:

1 Hazard and Operability Study[3]
This analytical method has been discussed above.

2 Technique of Operations Review[5]
This analytical technique or tracing system directs system designers and managers to examine the underlying and contributory factors that combine together to cause a failure of the system. It is associated with the theory of multicausality of accidents.

3 Gross Hazard Analysis
This analysis is done early in the design stage, and would be a part of a 'Hazop' (hazard and operability) study. It is the initial step in the system safety analysis, and it considers the total system.

4 Classification of Risks
This analysis involves the identification and evaluation of risks by type and impact (i.e. maximum potential loss) on the company. A further analysis – Risk Ranking – may then be undertaken.

5 Risk Ranking
A rank ordering of the identified and evaluated risks is drawn up, ranging from the most critical down to the least critical. This then enables priorities to be set, and resources to be allocated.

6 Failure Modes and Effects
The kinds of failures that could happen are examined, and their effects – in terms of maximum potential loss – are evaluated. Again this analysis would form part of an overall Hazop study.

7 Fault Tree Analysis
Fault tree analysis is an analytical technique that is used to trace the chronological progression of factors contributing to the accident situation, and is useful not only for system safety, but also in accident investigation. Again, the principle of multicausality is utilised in this type of analysis.

2.4 Systems theory and design

The word 'system' is defined in the Oxford Dictionary as 'a whole composed of parts in orderly arrangement according to some scheme or plan'. In present day parlance, we tend to think of 'systems' as connected with computers. However, the word is used in a wider sense in Operational Research to imply the building of conceptual and mathematical models to simulate problems and provide quantitative or qualitative information to executives who have to control operations, e.g. a maintenance system, a system governing purchase and use of protective clothing, a training system etc.

In this chapter, only an outline can be given of the concepts underlying systems theory and the theory will be presented mainly as an aid to clear thinking. The mathematical techniques associated with quantifying it can be found in textbooks of operational research.

The essential components of a systems model are goals or objectives, inputs, outputs, interactions between constituent parts of the system (e.g. storage, decision making, processing etc.) and feedback.

The stages in establishing and using a systems model are:

1 Define the problems clearly.
2 Build a systems diagram (including values).
3 Evaluate and test the system using already solved problems to check that the model gives the correct answer.
4 Use the model on new problems.

If we take as an example the provision of cost-effective machinery guards, we might produce a diagram such as *Figure 2.2* to indicate some of the factors affecting the process.

Such a conceptual type of model shows not only the sequence of events taking place, but further highlights feedback (fb on model) which informs management whether or not legal requirements are being satisfied. Besides the legal and technical considerations the diagram shows that the new guard could upset previously agreed incentive earnings and lead to conflict between management and unions, which in turn may lead to work stoppage and delays. The aim of the organisation (the system within which the subsystem is embedded) is to satisfy its customers and this can only be achieved by consistent output both in terms of quantity and quality. It can be seen that the fitting of a relatively insignificant machine guard can affect wider areas of the company's operations. Systems diagrams can direct the attention of those who are responsible for the effective running of a company to possible interactions, between either individuals or groups, inside or outside the company which could lead to conflicts and hence work disruption, long before it actually occurs, thus enabling suitable provisions to be made.

It is a useful exercise to consider how the safety adviser fits into the above system and which activities he should be involved in.

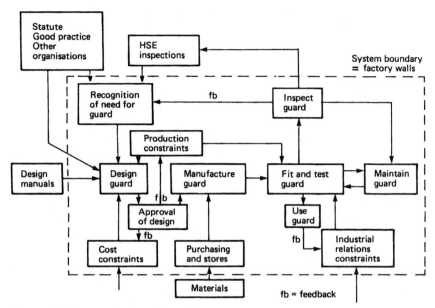

Figure 2.2 Systems diagram of the provision of a guard

The example shows a system boundary drawn to correspond to the boundary of the organisation. The fact that there are inputs and outputs across this boundary indicates that the system is an open one. Closed systems have no transactions across the system boundary. Consider a simple example, which compares these two main types. A steam engine's speed is controlled by a valve which controls the supply of steam. If the valve is adjusted by an attendant (an outside agent), the system is open, whereas if the valve is controlled by a governor responsive to the engine speed, the system is a closed one.

The system boundary could be drawn at various levels – e.g. in the guarding example it could be drawn at the level of the department in which the machine is located, the works, the company, or the country (in the last case there might be inputs across the boundary (frontier) of materials, designs or EC regulations which would still make it an open system). For a full systems analysis and model it is usually necessary to produce a hierarchy of diagrams showing the total system, main subsystems and subsubsystems etc.

System diagrams sometimes only contain the hardware or technical elements as in *Figure 2.3* of a simple diagram of a car.

This is a very incomplete system diagram as it leaves out human control. A complete model or sociotechnical system should include both technical and human aspects (both desired and undesired, e.g. vandalism, sabotage etc.). Only in this way can the mode of operation or breakdown of the whole system be investigated and if necessary redesigned. A useful exercise is to add the human element to the car

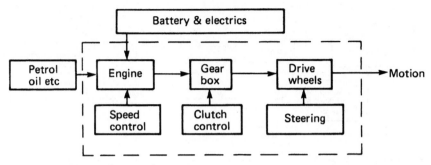

Figure 2.3 System diagram of a car

system above, or to devise a complete sociotechnical systems diagram for a company.

Accidents can be modelled as breakdowns of systems. The individual as a system set out in Chapter 5 is one example. Another which illustrates a fatal forklift truck accident in a warehouse is given in *Figure 2.4.*

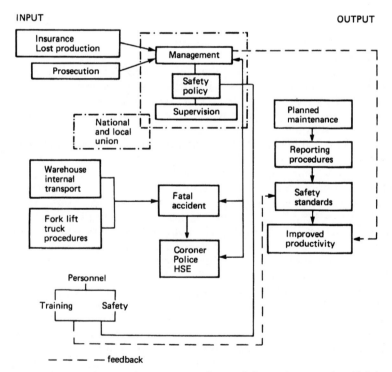

Figure 2.4 Systems diagram showing an accident and the environment in which it occurred

2.5 System safety engineering

System safety engineering has been defined[6] as an element of systems engineering involving the application of scientific and engineering principles for the timely identification of hazards and initiation of those actions necessary to prevent or control hazards within the system.

It draws upon professional knowledge and specialised skills in the mathematical, physical and related scientific disciplines, together with the principles and methods of engineering design and analysis to specify, predict, and evaluate the safety of the system.

An examination of system safety engineering methodology requires the consideration of two basic and interrelated aspects, namely system safety management and system safety analyses.

System safety management provides the framework wherein the findings and recommendations resulting from the application of system safety analysis techniques can be effectively reviewed and implemented.

System safety analyses employ the three basic elements of identification, evaluation, and communication to facilitate the establishment of cause. System safety analyses provide the loss identification, evaluation and communication factors and interactions within a given system which could cause inadvertent injury, death or material damage during any phase or activity associated with the given system's life-cycle.

Examples of system safety analyses include: routine hazard spotting; job safety analysis; hazard and operability studies; design safety analysis; fault-tree analysis; and stimulation exercises using a computer.

2.6 Fault tree analysis

2.6.1 Introduction

Fault tree analysis is a technique that may be utilised to trace back through the chronological progression of causes and effects that have contributed to a particular event, whether it be an accident (industrial safety) or failure (system safety).

The fault-tree is a logic diagram based on the principle of multi-causality that traces all the branches of events that could contribute to an accident or failure.

2.6.2 Methodology

In constructing a fault-tree to assist in cause analysis, firstly the event – the accident or failure – must be identified. Secondly, *all* the proximate causes (contributory factors) must be investigated and identified. Thirdly, each proximate cause (i.e. each branch of a contributory factor) must be traced back to identify and establish all the conceivable ways in which each might have occurred. Each contributing factor or cause thus identified is then studied further to determine how it could possibly have happened, and so on, until the beginning or source of the chain of events has been highlighted for each branch of the fault-tree.

Certain standard symbols are used in the construction of a fault-tree; some of the more common are:

The rectangle identifies an event or contributory factor that results from a combination of contributory factors through the logic gates.

The 'AND' logic gate describes the logical operations whereby the co-existence of *all* input contributory factors are required to produce the output event or contributory factor.

The 'OR' logic gate defines the situation whereby the output event or contributory factor will exist if *any* one of the input contributory factors is present.

Examples of fault-tree analyses are presented in Bird and Loftus[7] and Petersen[8].

By tracing back in this way the causes of accidents during accident investigation, a clearer and more objective assessment may be made of *all* the contributory factors, and hence more effective preventive action may be taken to ensure that there is no recurrence.

2.7 Probabilistic risk assessments

A probabilistic risk assessment consists of the following stages:

(a) the identification of undesired events and the mechanisms by which they occur (WHAT IF?),
(b) the likelihood – probability – that these undesired events may occur (HOW OFTEN?),
(c) the consequences of such an event once it occurs (HOW BIG?),
(d) a calculated judgement as to the significance of (b) and (c) (SO WHAT?) which may or may not lead to,
(e) the taking of control action.

Stage (a) – Identification of undesired events
Primarily, this stage involves the use of HAZOP studies (see section 2.1.2). Hazops are usually carried out at various stages of a system's cycle:

● conceptual design stage
● detailed design stage
● operational stage.

Stage (b) – Likelihood of undesired events
This stage involves an estimation of the probability of whether the undesired event is likely to occur or not.

Probabilities can be established mathematically, based upon the probable failure rates of individual components within the system. Data on individual components may be obtained from manufacturers' reliability statistics or quality assurance information. Specific failure rate data for individual items can also be obtained from reliability data banks such as that operated by the United Kingdom Atomic Energy Authority's (UKAEA) System Reliability Service.

Aspects such as maintenance schedules, condition monitoring, replacement criteria and human reliability/failure should also be taken into account.

An ideal technique for summating these individual probabilities to obtain the overall probability of the event occurring is Fault-Tree Analysis (see section 2.6), which is in essence a logic diagram with the event at the top of the tree.

Stage (c) – Consequences of undesired event occurring
Initially a hazard analysis (HAZAN) is undertaken so as to ascertain the magnitude of the potential problem and its potential for harm to the people, plant, process and the public.

A subsequent risk analysis will then go on to examine the actual consequences – worst possible case considerations – and express them in quantifiable terms. This then enables Stage (d) to be performed.

Stage (d) – Is the risk of the event occurring significant?
The output from Stage (c) may be expressed in the form of individual risk or of societal risk. Individual risk is the probability of death to an individual within a year (e.g. 1 in 10^4 per year). Societal risk is the probability of death to a group of people – either employees or members of the general public – within a year (e.g. a risk of 500 or more deaths of 10^{-8} per year).

Societal risks are usually given as fatal accident frequency rates (FAFRs).

The fatal accident frequency rate is defined[9] as the number of fatal injury accidents in a group of 1000 in a working lifetime (10^8 hours).

In making judgements to enable decisions concerning control action (Stage (e)) to be made, use is made of (published) risk criteria. These criteria are expressed in the form of numerical risk targets and they provide a yardstick for decision-makers against which to judge the significance of estimated risks. Generally two forms of risk criteria – as indicated above – are used:

● employee risks (on site)
● public risks (off site).

Stage (e) – Taking control action
If the calculated risk criteria figure is above the agreed accepted (published) figure, then control action is necessary. The amount by which

the calculated figure is higher than the agreed figure is useful in setting priorities, i.e. the greater the difference, the higher is the priority for control action.

If the calculated risk factor is below the accepted figure, then the safety provisions of the system may be considered to be adequate, and hence no further control action is required.

For employees in the UK chemical industry the FAFR is approximately 3.5. An acceptable target would obviously be below this, at approximately half, i.e. 1.7.

For the general public, FAFRs are rare. However, it has been suggested[10] that from an individual risk viewpoint as involuntary risks expose members of the public to a risk of death of about 10^{-7} per person per year, then industrial activities should not increase this figure. Hence, a risk criteria of less than 10^{-7} is acceptable.

2.8 Health and safety in design and planning

2.8.1 Introduction

The consideration of health and safety aspects at the design and planning stages of new projects, buildings, plant and processes is vitally important in order to ensure that health and safety are built in, rather than bolted on.

It is therefore essential that all engineers, designers, and architects receive education and training in such matters, so as to ensure that relevant legislative and technical factors appertaining to health and safety are taken into account at the design and planning stage of all new projects.

Certain risk identification (e.g. Hazop) and risk evaluation techniques (e.g. Hazan, fault-tree analysis) that may prove useful in this regard have been discussed above.

2.8.2 Project design

It is imperative that assessment and control of all new projects take health and safety aspects into account at the earliest – and at all – stages of a project's development.

The project originator should ensure that the project is appraised from a health and safety viewpoint, and it should not be allowed to proceed until it has been approved by a safety adviser. Ideally the safety adviser should be involved at all stages of the project's design and planning, so that specialist guidance and advice may be incorporated as necessary.

The risks associated with new projects may include: use of hazardous substances; insufficient product data; faulty electrical equipment; poor access/egress; poor ergonomics; noisy equipment; poorly guarded machinery; imported equipment/materials; lack of risk assessment; lack of training/awareness on behalf of management, supervision and employees; poor environmental control; inadequate emergency procedures; inadequate maintenance considerations; poor construction methods; little or no consideration of waste disposal/demolition.

The whole life-cycle of the project – from inception to ultimate disposal – should be considered at the design stage.

2.8.3 Project design: health and safety action plan

When health and safety is considered at the design stage, the following action plan should ensure that risks are designed and engineered out of the system *before* they are able to cause injury, disease, damage or loss:

- Ensure advice on health and safety is made available to the project team/ originator.
- Ensure a Hazop-type brainstorming meeting of key personnel associated with the development of the project is held to identify risks and establish control actions. (The list of risks in section 2.8.2 may serve as a checklist.)
- Ensure that suitable written safe systems of work are prepared and communicated to all concerned – i.e. develop the 'software' to go with the 'hardware'.
- Ensure that all aspects of the project comply with relevant legislative and technical standards.
- Ensure that all personnel concerned with the project receive necessary health and safety training.
- Ensure that suitable emergency procedures are developed.
- Ensure that the project commissioning procedure involves approval by the safety and health adviser at all stages of the project's development.

2.8.4 Project commissioning

Once the project has been approved, practical aspects of supply, installation, commissioning, use and ultimately disposal follow on. As stated in section 2.8.2, health and safety should already have been considered at the design stage of the project.

From a legislative viewpoint, the supply, installation and commissioning aspects are covered by s. 6 of the Health and Safety at Work etc. Act 1974, which requires manufacturers, suppliers, installers etc. of articles and substances to ensure that they are safe and without risks to health when set, used, cleaned maintained, stored, transported and disposed of.

In order to ensure safe commissioning of new plant and equipment, a three-part plant acceptance system should be utilised:

Part one   – provisional safety certificate (for test purposes)
                 – this enables only design/engineering personnel to undertake testing, once approved by safety adviser and project originator.

Part two   – trial production run
           – this enables production employees to become familiar with the new equipment under the supervision of the project originator/safety adviser and enables any previously unforeseen risks to be engineered out at the man/machine interface.
Part three – final certification
           – once all testing and production trials have been satisfactorily completed, the plant/equipment is handed over to production management.

By involving a multidisciplinary team – including a safety adviser – at all stages of the design, planning and commissioning process, the risk of having to provide additional – and more costly – safeguards *after* the plant is in full use is minimised.

2.9 Quality, Environment, Safety and Health Management Systems (QUENSH)

2.9.1 Introduction

Established quality assurance procedures provide a sound basis for the development of systems for health and safety management[11]. The introduction of the HSE's publication *Successful Health and Safety Management* (revised 1997) clearly states[19] that many of the features of effective health and safety management are indistinguishable from the sound management practices advocated by proponents of quality and business excellence.

This logic has more recently been extended into the realms of environmental management systems via BS 7750: 1994[25]/BS EN ISO 1400: 1996[26] and also into occupational health and safety management systems via BS 8800: 1996[27]. Indeed, Annex A to BS 8800: 1996 clearly demonstrates the commonality of approach between BS 8800 and BS EN ISO 9001: 1994[28]. for quality management systems. Also BS 8800 postulates two routes towards occupational health and safety management systems: route 1 via HS(G)65[19]; route 2 via BS EN ISO 14001: 1996[26].

Hence the QUENSH approach towards integration is gathering momentum.[29,30]

2.9.2 Quality systems

The original British Standard on quality systems[12] – BS 5750: 1987 – has now been superseded by BS EN ISO 9001: 1994[28]. This tells suppliers and manufacturers what is required of a quality-orientated system from a practical viewpoint. It identifies basic disciplines and specifies procedures and criteria to ensure customer requirements.

Within the context of ISO 9001, quality means that the product is fit for the purpose for which it has been purchased, and has been designed and constructed to satisfy the customer's needs.

The Standard sets out how an effective and economic quality system can be established, documented and maintained.

The Standard considers that an effective quality system should comprise: management responsibility; quality system principles; quality system audits; quality/cost considerations; raw material quality control; inspection and testing; control of non-conforming product; handling, storage, packaging and delivery; after sales service; quality documentation and records; personnel training; product safety and liability; and statistical data/analyses.

2.9.3 Quality and safety

Although the Standard does not explicitly refer to 'people safety', there are obvious parallels to be drawn between the quality systems approach and health and safety management.

Indeed, the management systems described in ISO 9001 are as applicable to health and safety management as they are to product liability risk management[11]. ISO 9001 is concerned with the achievement of quality, which is measurable against specific criteria. It lays down systems which demonstrate achievement against these specified criteria.

One of the benefits of an effective quality system is to minimise the risk of product liability claims and losses.

In the case of product liability risk management, the specified criteria of performance are:

- Compliance with s. 6 of the Health and Safety at Work etc. Act 1974 (as modified by the Consumer Protection Act 1987).
- Compliance with all other relevant statutory provisions, especially the Management of Health and Safety at Work Regulations 1992.
- The ability to adhere to all product contract conditions.
- The minimisation of defective products.
- The maximisation of health and safety benefits to the consumer.

This parallels very closely the perceived criteria of an effective health and safety management system, namely:

- Compliance with all aspects of the Health and Safety at Work etc. Act 1974.
- Compliance with other relevant statutory provisions – e.g. Regulations and those parts of the Factories Act 1961 that remain in effect.
- The ability to adhere to the common law duty of care, and relevant aspects of employment contract conditions.
- The minimisation of risks likely to cause injury or disease.
- The maximisation of health and safety benefits to employees, third parties, and the general public.

From the above, it may be seen that the application of quality systems to the management of health and safety at work has distinct benefits, especially when consideration is given to the tremendous overlap between the two subject areas. Overlap examples include: policies; systems and procedures; standards; documentation – records; training (including record keeping); statistical analyses – causal, numerical; accident/complaint investigations; audits/inspections (internal and external); and the taking of remedial control action.

Hence effective quality systems management will greatly enhance the management of health and safety, and will lead to an overall improvement in the level of safety performance.

2.10 Use of data on accidents

In addition to the use of qualitative and quantitative accident data when identifying and evaluating risk (see sections 2.1.2 and 2.1.3), there is a need to consider the relative occurrence of different accident types, in order that effective accident control measures may be implemented throughout an organisation.

Accidents, whether they result in injury, damage, disease or loss, need to be controlled. Similarly those that have no end result – i.e. the near-misses – should be considered for control action.

To enable an accident control system to be developed, it is necessary that *all* accidents are reported, recorded, investigated and analysed, so that after remedial action has been decided, plans can be drawn up to prevent a recurrence. The most important question to be asked in any accident investigation is: 'What action has been taken to prevent a recurrence?'

The collection of accident data on a much broader base to facilitate the planning of control action has been undertaken by a number of researchers and one of the most widely applied accident ratios is that propounded by Bird[13] in 1969.

The Bird study is generally depicted in triangular form:

From this it can be seen that the majority are either 'near-misses' or damage only accidents, so that any accident control programme, for its greatest effect, must concentrate much of its attention on these two areas. Related techniques of damage control and incident recall are dealt with in sections 1.4.2 and 2.12 and also by Carter[14].

Accident data are only one form of risk information which assists in the identification, assessment and control of risks. Other forms of risk information include: reliability data; epidemiological studies; mortality/ morbidity data; vulnerability analysis; results of audits/inspections; insurance claims data; and – probably the best – personal experience. Sources of risk information are listed in the Further Reading section of this chapter (see especially the *Handbook of Risk Management* by Carter *et al.*).

2.11 Maintenance systems and planned maintenance

2.11.1 Maintenance – risk management aspects

From a risk management viewpoint, health and safety aspects concerning all maintenance operatives should be considered and planned for at the design stage of a new project or process.

Aspects such as: safe access; operations at unguarded machinery; breaking into pressurised systems; blanking off; inerting; hot work; installation; setting up; inching; dismantling; and demolition, should have been taken into account.

Effective maintenance systems that take into account health and safety aspects serve not only to increase the lifetime of key plant and equipment, but also ensure safer and more cost-efficient operation.

Indeed, a planned maintenance system is essential to the continued safe operation of a range of plant and equipment, as well as meeting both general and specific legislative requirements – e.g. statutory inspections.

Plant and equipment requiring such regular inspection and planned maintenance include: boilers; pressure vessels – air receivers; power presses; cranes; lifting equipment – chains, pulleys; ventilation systems; pressurised systems; access equipment – ladders; scaffolding; electrical equipment – portable electric tools; office equipment; emergency eye-wash points and showers; fire alarm systems and fire points; fire extinguishers; fixed firefighting appliances/systems; stairways, walkways, gangways; in addition to all key items of production plant and equipment.

2.11.2 Planned maintenance

It is generally accepted that a system which only allows for the maintenance/repair of plant on a crisis or breakdown basis is not cost-efficient, especially in the longer term.

Hence any form of planned maintenance – whether it be planned lubrication at one end of the scale, or sophisticated techniques such as condition monitoring at the other end – will improve both safety and plant integrity/reliability.

A system of planned maintenance may be introduced progressively, commencing with limited maintenance routines only on key items of plant and equipment. This may then be extended to a wider range of plant, utilising more complex routines, until such time as a 100% planned maintenance system has been implemented. As with quality systems, maintenance records and histories are essential to ensure smooth operation.

When a total (100%) system of planned maintenance has been implemented, then, by definition, any breakdown that occurs must be accidental in nature. Hence such accidental breakdowns – i.e. damage accidents – should be reported and investigated (see section 2.12 for fuller discussion).

2.12 Damage control

2.12.1 Introduction

The technique of damage control involves the systematic reporting, investigation, costing and control of damage accidents within an organisation.

2.12.2 Reporting

The reporting system for damage accidents within an organisation should, as far as possible, be the same as that for injury accidents. The fundamental differences between injury and damage accidents are that in the former, the victim (the injured person) starts the reporting chain by advising his supervisor or first-aider of his injury.

However, in the case of damage accidents, the victim is an inanimate object, physically incapable of reporting the accident. Therefore, there is a need to develop a system by which accidental damage can be highlighted and clearly identified.

It would appear that the situations likely to give rise to accidents can be divided into three main categories:

(i) Fair wear and tear – if we define this as the failure of an item in service after a lifetime of proper use, then, in theory, such a failure will be eliminated by a total (100%) system of planned maintenance and condition monitoring. In such a situation any failure will be an accident, i.e. unexpected, unplanned.

(ii) Malicious damage – this will need to be identified and dealt with separately.

(iii) Accident damage – it is this final category that we are dealing with in this section.

There will, of course, be many occasions in practice where it will be necessary to determine whether damage arises from (i) or (iii) above and the opinion of all relevant staff will need to be sought.

2.12.3 Investigation

In order to ensure that prompt action is taken to prevent the damage accident recurring, a detailed investigation should be undertaken. This should aim to establish the sequence of events, and to identify the combination of causes of the damage accident.

It should be remembered that the prime aim of any accident investigation is prevention, rather than blame. As with injury accidents, the investigating team should generally comprise the relevant supervisor and/or manager, together with the safety adviser and safety representative (if appointed).

To aid the investigation it is suggested that a detailed accident investigation report form is utilised.

2.12.4 Costing

The prime aim in costing damage accidents is to obtain an objective measure of severity. This will in turn assist in the setting of priorities and hence the allocation of resources and the overall planning of the damage control programme.

Information regarding the cost of all accidents – both injury and damage – may also be used to motivate management action in accident prevention.

2.12.5 Control

During the investigation the question 'What action should be taken to prevent a recurrence?' should be clearly answered. This should ensure that the correct remedial action is promptly taken.

The findings of the investigation should be clearly communicated to the manager responsible for instigating the control action.

To gain the full value from a control system, it is important that findings be communicated to all management within the organisation who may be faced with similar problems. This will greatly assist in the elimination of the causes of potential accidents before either injury or damage results.

2.13 Cost-effectiveness of risk management

2.13.1 Accident costing

The three fundamental arguments that may be employed to promote action in the field of risk management concern the legal, humanitarian, and economic aspects (see section 1.7).

For the vast majority of organisations, there will be few or no quantitative data relating to the economic facets, and particularly to the costs associated with accidents. In order to be able to demonstrate cost-effectiveness (or cost benefit), there is therefore a need to be able to quantify the cost of all losses associated with accidents.

An investigation of the costs of occupational accidents for the UK as a whole[15] has considered the costs under two headings: resource costs covering lost output, damage to plant, medical treatment and administrative costs, and subjective costs relating to pain and suffering of the victim and his family.

	Resource costs £	Subjective costs £	Total costs £
Fatality	71 500	38 000	109 500
Serious injury	1 800	2 750	4 550
Temporary disabling diseases	1 600	2 750	4 350
Minor injuries	300	150	450

However figures relating to the national situation are often not pertinent to individual factories and departments. Some figures obtained in 1998 for the uninsured costs of three types of accident[16] were:

Lost time injury	£1976
Non-lost time injury	£ 31
Damage	£ 133

The above figures – in isolation – do not have much impact, but must be related to the total costs of accidents in an individual factory or organisation. When these accident costs are compared with other business costs, such as production, sales and distribution, a true indication of the drain on financial resources can be appreciated. The following hypothetical case is presented as an example:

Consider Factory Z in 1997; the data given below apply:

Number of lost time injury accidents	=	30
Number of non-lost time injury accidents	=	750
Number of damage accidents (estimated)	=	780
Employers' liability premium (£1.10% wages)	=	£90 000
Number of employees	=	1000

When the average uninsured accident cost figures are applied to the above data, the following result:

Uninsured cost of lost time injury accidents
$$30 \times £1976 \quad = \quad £59\,280$$
Uninsured cost of non-lost time injury accidents
$$750 \times £31 \quad = \quad £23\,250$$
Uninsured cost of damage accidents $780 \times £133$ =   £103 740
Total uninsured accident cost   =   £186 270

Adding to this the Employer's Liability insurance premium (£90 000), the total accident cost becomes £276 270, or approximately £276 per employee per year.

Of particular concern is the impact of such costs on the overall profitability of the organisation. Although the total cost of accidents within a company may be relatively small, it can amount to approximately 2% of the annual running costs, and represents a direct drain on profits.

The table below illustrates the sales necessary to cover these accident costs.

Accident costs £	If your organisation profit margin is		
	1%	3%	5%
1 000	100 000	33 000	20 000
10 000	1 000 000	330 000	200 000
100 000	10 000 000	3 300 000	2 000 000

Any reduction of these costs that may be made through a cost-effective risk management programme will lead both to a safer and more profitable organisation.

The use of cost figures presented in this way will be more meaningful to managers and executives and is likely to stimulate their motivation to reduce the number of accidents.

An alternative – more specific – method of relating accident and wastage costs to costs of production is presented in the following example:

On a construction site, the number of facing bricks lying around was estimated by random counting to be 1300, valued at £455.

The cost in terms of profit on turnover (of 6.3%) was:

$$\frac{\text{Cost} \times 100}{\text{Profit on turnover}} = \frac{455 \times 100}{6.3\%} = £7222$$

i.e. a turnover of £7222 was required to pay for the bricks – i.e. to break even.

The cost expressed as a percentage of the total contract price (of £1 650 000) was:

$$\frac{\text{Turnover to pay for bricks} \times 100}{\text{Contract price}} = \frac{£7220 \times 100}{£1\ 650\ 000} = 0.43\%$$

Such exercises may be undertaken for all types of accident and wastage situations, to enable the costs involved to be judged in relative – rather than absolute – terms.

2.13.2 Cost benefit analysis

Cost benefit analysis techniques have been developed in recent years, as decisions concerning risk management have been made on a cost versus risk basis – i.e. so far as is reasonably practicable.

Indeed, all proposed legislation has to pass the cost/benefit test at the consultative stage, before being allowed to pass on to the statute books. Although, in the majority of cases, the benefits of proposed legislation are not accurately quantified, nevertheless the qualitative benefits are listed, and these may be seen to outweigh the costs. Further discussion on the cost/benefit of specific proposed legislation may be found in relevant Health and Safety Commission Consultative Documents which contain a review of the costs and benefits associated with proposed changes in legal requirements.

To undertake a cost benefit analysis, answers to the following questions are required:

- What costs are involved to reduce or eliminate the risk?
- What degree of capital expenditure is required?
- What ongoing costs will be involved, e.g. regular maintenance, training?
- What will the benefits be?
- What is the pay-back period?
- Is there any other more cost-effective method of reducing the risk?

The cost factors associated with poor risk management have been discussed in sections 1.7 and 2.13.1 and include both insured and uninsured elements.

The benefit factors should initially be listed, and should always be quantified, where possible, so that the pay-back period can be established. Some benefits are easier to quantify than others.

Benefits of effective risk management include:

- few claims resulting in lower insurance premiums,
- less absenteeism,
- fewer injury and damage accidents,
- better levels of health,
- higher productivity/efficiency,
- better utilisation of plant and equipment,
- higher morale and motivation of employees,
- reduction in cost factors.

The costs are then balanced against the benefits (both qualitative and quantitative) and then an objective decision may be made on whether to allocate resources to the project or not. This will usually be based on the length of the pay-back period. Most health and safety projects will generally have a pay-back period of between three and five years – i.e. medium-term rather than short term.

2.14 Performance evaluation and appraisal

2.14.1 Introduction

In the vast majority of company health and safety policies the health and safety responsibilities of line and functional management are clearly laid down, together with – in some cases – the mechanism by which the fulfilment of these responsibilities will be monitored.

Indeed, the Accident Prevention Advisory Unit (APAU) have produced three excellent publications[17-19] which provide additional guidance and discussion on the aspects of policy management, implementation and monitoring.

2.14.2 Financial accountability and motivational theory

However, the most effective way to fix accountability for health, safety and indeed risk management responsibilities is by financial accountability of directors and managers.

This is borne out when consideration is given to the use of the legal, humanitarian and economic arguments for health and safety at work (section 1.7).

Maslow[20] related his theory of motivation to human needs. He suggested five sets of goals which are usually depicted as a progression or hierarchy:

Self actualisation (Achievement; Doing a good job)
↑
Esteem (Status; Approval)
↑
Love (Social)
↑
Safety (Security)
↑
Physiological (Sustenance)

The logic is that once the lower needs are well satisfied, the individual is motivated to attain satisfaction at the next higher level, and so on up the hierarchy.

From a health and safety viewpoint, therefore, the humanitarian argument tends to operate in terms of doing a good job, caring for people, and being well thought of and accepted, i.e. the higher end of the hierarchy: esteem/self actualisation.

The legal argument tends to motivate via the safety (security) need – the real (if remote) threat to physical security – and the esteem (approval) need of others. This motivation is in the middle of the hierarchy.

However, the economic argument derives its motivational impact from the fact that the goals involved tend to be generally lower in the hierarchy – i.e. the safety (security) need and, in certain cases, the physiological need. The security need involves the manager's performance in his job –

the ability to effectively carry out his responsibilities whilst keeping within budgetary constraints. Hence adverse performance could result in a decrease in or loss of the next salary increment or merit rise, as part of the performance appraisal exercise, thus directly threatening the security needs. In times of economic recession, this failure to achieve satisfactory performance could even affect the manager's position in the company and his ability to maintain employment, thus threatening the physiological needs.

Hence it would appear that financial accountability is the key.

2.14.3 Use of accident costs

However, the present system of accounting for health, safety, accident prevention and risk management operating in most of industry does not attempt to make line management financially accountable for accidents and uninsured losses, and very little use is made of economic arguments in stimulating management interest in risk management. Any arguments put forward rely mainly on legal and humanitarian considerations which, in some instances, fail to convince management that there is a need for risk management, at least beyond compliance with statutory duties.

One method that may prove useful involves the use of budgetary control which would introduce economic accountability into the field of accident prevention. Such measures would involve the reorganisation of the existing accounting procedures in most companies in order to overcome the lack of accountability for accident prevention and risk management.

When an accident occurs within a factory department, the cost of the accident usually is absorbed into the running costs of the factory as a whole, and will not be itemised on the departmental balance sheet. Nor will many of the uninsured costs be paid for from the departmental manager's budget. Furthermore, the insured cost – i.e. insurance premiums, such as employer's liability, will generally be paid from a central fund, usually administered by Head Office.

However, when the company safety adviser or factory inspector recommends safety measures such as guarding for machinery, the cost is usually charged against the departmental manager's budget, though it is very unlikely that it would be itemised as an accident prevention cost.

Thus under accounting systems currently employed in many companies, accident costs are not charged to departmental managers' budgets, whereas accident prevention is. Hence the departmental manager has no economic motivation to undertake any accident prevention; rather the reverse.

A positive economic motivating factor for encouraging accident prevention may be introduced by interchanging the budgetary system. For each accident – injury or damage – that occurs within a department, a charge is made against that department. Any accident prevention expenditure that is required within the department is financed from a central fund subject to approval by the risk manager or safety adviser. The result – as far as the departmental manager is concerned – is that it is costly to have accidents but not to prevent them.

Thus line management become accountable for the accidents occurring within their areas of control. At the end of the financial year, when the budgets are drawn up, a realistic allowance for accidents will be set within the budget as a target for the manager to achieve. Failure to achieve the agreed target would adversely reflect on performance and should be taken into account at job performance appraisal interviews. This allowance will form an integral part of the management plan as with budgetary control in other business areas.

However, the number – and hence the cost – of accidents budgeted for should be less than the previous year so that reduction of accidents becomes part of the management plan. The reorganised system would bring about the necessary economic accountability and would make full use of the knowledge and data obtained in establishing what accidents were costing the company in financial terms.

Once the costs of accidents have been established, the reorganised budgetary system can be implemented. The charges to be made against the departmental manager's budget can then be calculated and allocated on a monthly basis. The departmental manager might receive a monthly report giving information on the costs of accidents and accident prevention expenditure. This would enable him to plan any action necessary to maintain or improve the level of safety within his department. Also it would facilitate decision-making in connection with the allocation of scarce resources. Any deficiencies in the current programme would be highlighted in cost terms rather than by a frequency rate – a measure of safety increasingly questioned by safety advisers.

On their own, the legal and humanitarian arguments for risk management may not be sufficient to achieve a reduction in accidents and other losses. The addition of economic accountability – through accident costing – should greatly assist in reducing losses arising from accidents and ill-health at work.

2.15 Loss control profiling

General aspects of loss control are discussed in section 1.4. Loss control profiling is one of the major evaluation and control techniques associated with loss control management. The technique of profiling has formed the basis for a number of proprietary auditing systems such as International Safety Rating System (ISRS), Complete Health and Safety Evaluation (CHASE) and Coursafe.

Between 1968 and 1971, Bird[21,22] designed a loss control profile to quantify management's efforts in this area. He considered 30 areas of management activity that are connected either directly or indirectly with the reduction of loss.

These 30 areas are:

1 Management involvement and policy making.
2 Professional competence of loss control manager.
3 Technical experience of loss control manager.

4 Aptitude and talents of loss control manager.
5 In-depth accident investigations.
6 Plant and facility inspection.
7 Laws, policies, standards.
8 Management group meetings.
9 Safety committee meetings.
10 General promotion through the use of posters, banners, signs.
11 Personal protection.
12 Supervisory training.
13 Employee training.
14 Selection and employment procedures.
15 Reference library.
16 Occupational health and hygiene.
17 Fire prevention and loss control.
18 Damage control.
19 Personal communications.
20 Job safety analysis.
21 Job safety observations.
22 Records and statistics.
23 Emergency care and first aid.
24 Product liability.
25 Off-the-job safety including on the road and at home.
26 Incident recall and analysis.
27 Transport including managers and salesmen driving cars.
28 Security.
29 Ergonomic applications.
30 Pollution and disaster control.

Each of the 30 areas needs to be evaluated to pinpoint where action is necessary to improve the organisation's control of losses. The evaluation should be undertaken by trained personnel using the technique of asking a series of questions related to each of the 30 areas. Up to 500 questions may be required to cover the 30 areas.

Thus, the first stage in loss control profiling is to develop the list of questions that relate specifically to the organisation under review.

The answer to each question is rated on a scale ranging from 0% for a bad to 100% for a good response. An evaluation of each of the 30 areas is then calculated by taking the average percentage of those answers relating to a particular area.

A percentage figure of 25% or less indicates those areas where immediate action needs to be taken. A percentage figure between 25% and 50% indicates those areas where there is a need for improvement within the near future. A percentage figure between 50% and 75% indicates those areas in which an acceptable level has been achieved, but in which there is still room for improvement. A percentage figure of 75% or more indicates those areas where the organisation is operating at optimum performance, but which have to be monitored to ensure that this performance is maintained.

The results of such evaluations can be presented graphically in the form of a horizontal bar chart where each subject area is shown on a separate line. Those areas giving cause for concern, i.e. the short lines, are immediately highlighted.

Fletcher and Douglas[23] and Fletcher[24] developed Bird's original ideas on profiling and formulated their own detailed evaluation questionnaire in which the answer to each question was rated on a six-point scale, ranging from 'fully implemented and fully effective' (score 5) to 'nothing done to date' (score 0). The scores of each question in a subject area are then summated, and the value is expressed as a percentage of the maximum attainable score. Loss control profiles are then constructed and utilised in a similar manner to that described above.

Once the losses – both actual and potential – have been evaluated, and a loss control profile developed, then – and only then – can a definite action programme of loss control be planned and implemented.

This would be based on assessing the deficiencies highlighted by the loss control evaluation and profile, then initiating a programme of work to make good those deficiencies.

Annual profiles may be undertaken to assess progress made, and also to ensure that all areas under review are maintained at an acceptable level.

References

1. Bird, F. E., and O'Shell, H. E., Incident recall, *National Safety News*, **100**, No. 4, 58–60 (1969)
2. Bamber, L., Incident recall – a (lack of) progress report, *Health and Safety at Work*, **2**, No. 9, 83 (1980)
3. Chemical Industries Association Ltd, *A Guide to Hazard and Operability Studies*, Chemical Industries Association Ltd, London (1977)
4. Bird, F. E. and Loftus, R. G., *Loss Control Management*, 464, Institute Press, Loganville, Georgia (1976)
5. Ref. 4, pp. 165–171
6. Ref. 4, p. 474
7. Ref. 4, p. 493 et seq.
8. Petersen, D. C., *Techniques of Safety Management*, 2nd edn, p. 174 et seq., McGraw-Hill, Kogakusha, USA (1978)
9. Kletz, T. A., *Hazop and Hazan: notes on the identification and assessment of hazards*, Institution of Chemical Engineers, Rugby (1983)
10. Kletz, T. A., Hazard analysis – its application to risks to the public at large (Part 1), *Occupational Safety & Health*, **7**, 10 (1977)
11. Industrial Relation Services, A systems approach to health and safety management, *Health & Safety Information Bulletin* No. 168, 5–6, Industrial Relations Services, London (1989)
12. BS 5750: 1987, *Quality systems*, British Standards Institution, London (1982)
13. Bird, F. E, *Management Guide to Loss Control*, 17, Institute Press, Atlanta, Georgia, (1974)
14. Carter, R. L., The use of non-injury accidents in risk identification, 4.6–01–4.6–05, *Handbook of Risk Management*, Kluwer Publishing Ltd, Kingston-upon-Thames (1992)
15. Morgan, P. and Davies, N., Cost of occupational accidents and diseases in GB, *Employment Gazette*, 477–485, HMSO (Nov. 1981)
16. Ref. 14, pp. 6.4–01–6.4–12
17. Health and Safety Executive, *Managing Safety, Occasional Paper Series No. OP3*, HSE Books, Sudbury (1981)
18. Health and Safety Executive, *Monitoring Safety, Occasional Paper Series No. OP9*, HSE Books, Sudbury (1985)
19. Health and Safety Executive, Publication No. H5(G)65, *Successful Health and Safety Management* (2nd edn.), HSE Books, Sudbury (1997)

20. Maslow, A. H., A theory of human motivation, *Psychological Review*, **50**, 370–396 (1943)
21. Ref. 13, pp. 151–165
22. Ref. 4, pp. 185–197
23. Fletcher, J. A. and Douglas, H. M., *Total Loss Control*, 113–154, Associated Business Programmes, London (1971)
24. Fletcher, J. A., *The Industrial Environment – Total Loss Control*, 18–122, National Profile Ltd, Willowdale, Ontario, (1972)
25. British Standards Institution, BS 7750: 1994, *Specification for environmental management systems*, BSI, London (1994)
26. British Standards Institution, BS EN ISO 14001: 1996, *Environmental management systems – Specification with guidance for use*, BSI, London (1996)
27. British Standards Institution, BS 8800: 1996, *Guide to occupational health and safety management systems*, BSI, London (1996)
28. British Standards Institution, BS EN ISO 9001: 1994, *Quality systems – Specification for design/development, production, installation and servicing*, BSI, London (1994)
29. Fishwick, L. and Bamber, L., Common Ground – Practical ways of integrating the environment into your health and safety programme (Part 1), *Health and Safety at Work 1996*, **18**, 2, pp. 12–14
30. Fishwick, L. and Bamber, L., Common Ground – Practical ways of integrating the environment into your health and safety programme (Part 2), *Health and Safety at Work 1996*, **18**, 3, pp. 34/35

Further reading

Diekemper and Spartz, A quantitative and qualitative measurement of industrial safety activities, *J. Amer. Soc. Safety Engrs*, **15**, No. 12, 12–19 (1970)

Fine, W. T., Mathematical evaluation for controlling; hazards, *J. Safety Research*, **3**, No. 4, 57–166 (1971)

Chemical Industries Association Ltd and Chemical Industry Safety and Health Council, *A Guide to Hazard and Operability Studies*, Chemical Industries Association Ltd, London (1977)

Petersen, D.C., *Techniques of Safety Management*, 2nd edn, McGraw-Hill, Kogakusha, USA (1978)

Bird, F.E. and Loftus, R.G., *Loss Control Management*, Institute Press, Loganville, Georgia (1976)

Heinrich, H.W., Petersen, D. and Roos, N., *Industrial Accident Prevention – A Safety Management Approach*, 5th Edn, McGraw-Hill, New York (1980)

Dewis, M. *et al.*, *Product Liability*, Heinemann, London, (1980)

Carter, R.L. *et al.*, *Handbook of Risk Management*, Kluwer Publishing Ltd, Kingston-upon-Thames (1992)

Health and Safety Executive, *Report – Canvey: an investigation of potential hazards from operations in the Canvey Island/Thurrock area*, HSE Books, Sudbury (1978)

Health and Safety Executive, Advisory Committee on the Safety of Nuclear Installations (ACSNI), ACSNI Study Group on Human Factors, *2nd Report – Human Assessment: A Critical Overview*, HSE Books, Sudbury (1991)

Health and Safety Executive, Publication No. H5(G)65, *Successful Health and Safety Management* (2nd edn.), MSE Books, Sudbury (1997)

Chemical Industries Association, *Guidance on Safety, Occupational Health and Environmental Protection Auditing*, Chemical Industries Association, London (1991)

Health and Safety Commission, *Management of Health and Safety at Work, Approved Code of Practice: Management of Health and Safety at Work Regulations 1992*, HSE Books, Sudbury (1992)

Chapter 3

The collection and use of accident and incident data

Dr A. J. Boyle

3.1 Introduction

Although the title of this chapter refers to accidents and incidents, there is no general agreement about how accidents and incidents should be defined. For this reason, the chapter begins with a discussion of the various types of data which might be included in these two categories and the practical implications of this discussion.

The remainder of the chapter is divided into five main sections:

1 The collection of accident and incident data and the systems which have to be in place if accident and incident data are to be collected and recorded effectively.
2 The UK legal requirements to notify accidents and incidents of particular types and levels of severity.
3 The main uses of accident and incident data such as the techniques for learning from the analysis of aggregated accident and incident data, using trend analysis, comparisons of accident and incident data, and epidemiological analyses.
4 Lessons to be learnt from individual accidents and incidents by means of effective investigations.
5 The use of computers with accident and incident data.

3.2 Types of accident and incident data

A commonly used distinction between accidents and incidents is that accidents have a specific outcome, for example injuries or damage, while incidents have no outcome of this type, but could have had in slightly different circumstances. The HSE uses the following definitions[1]:

- *Accident* includes any undesired circumstances which give rise to ill-health or injury; damage to property, plant, products or the environment; production losses or increased liabilities.
- *Incident* includes all undesired circumstances and 'near misses' which could cause accidents.

However, it is preferable to think of accidents and incidents as part of a single, much larger, group of undesired events or circumstances which varies in two main dimensions.

1 Qualitative differences of actual and potential outcomes, for example injuries, ill-health and damage.
2 Quantitative differences in outcomes, for example 'minor' injuries, 'major' injuries and damage.

Each of these dimensions will now be considered in more detail.

3.2.1 Qualitative differences

Table 3.1 shows some of the main categories of outcome and examples of each.

To a certain extent, the divisions shown in *Table 3.1* are artificial and some examples of overlaps are given below:

● A customer complaint may be about product safety.
● A spillage may result in injury, ill-health, and asset damage.
● An injury accident can also involve asset damage and damage to the environment.
● Some injuries, such as back injury, can result in chronic illness.

However, despite these divisions being artificial, it is traditional to keep them separate, with different specialists dealing with particular types of outcome. This chapter will continue this tradition by restricting discussion to the following categories:

Table 3.1. The main types of accident and incident data

Quality	Environment	Injuries	Health	Asset damage and other losses
Customer complaints	Spillages	Injuries to employees at work	Sickness absence	Damage to organisation's assets
Product non-conformances	Emissions above consent levels	Injuries to others at work	Chronic illness	Damage to other people's assets
Service non-conformances	Discharges above consent levels	Injuries during travel	Sensitisation	Interruptions to production
		Injuries at home		Damage arising from unsafe products
		Injuries arising from unsafe products		Losses from theft and vandalism

- Damage to people, including mental and physical damage, damage which occurs instantaneously (mostly injuries) and damage which is caused over a longer period of time (mainly ill-health).
- Damage to assets, including assets of the organisation, and assets belonging to other people which are damaged by the organisation's activities.

However, the principles described apply equally well to all categories and there should be little difficulty in generalising them if required.

There is one further point on qualitative differences. The preceding discussion has assumed that there is either an outcome (accidents) or no outcome (incidents). However, there is a type of outcome which does not fit either of these categories and that is the creation of a hazard. For example, people can quite safely lay cables across a walkway, but they have then created a hazard for themselves and others. It is not usual to deal with this type of outcome as part of a discussion of accidents and incidents but there appears no good reason for this exclusion and hazard creation will be discussed at relevant points in this chapter.

3.2.2 Quantitative differences

It is well known that accidents vary in severity, ranging from minor injuries through major injuries and ill-health to fatalities and catastrophic damage.

There is a relationship between the severity of the outcome and the frequency of the outcome. As the seriousness of the outcome increases, the frequency of that outcome decreases. This means that there are many more minor injuries, and cases of minor ill-health, than there are fatalities. In addition, there are many more 'near misses' than there are minor injuries or cases of minor ill-health. The sort of relationship which exists between frequency and severity is illustrated in *Figure 3.1.*

The relative numbers in this sort of relationship are not important, what is important is that it is recognised that there is a continuum from near miss to fatality and that definitions such 'minor', 'three day' and 'major' are arbitrary points on this continuum.

There have been various studies which have put numbers to the different categories of outcome and these are usually referred to as 'accident triangles'. A typical accident triangle is shown in *Figure 3.2.*

The figures given in *Figure 3.2* are from a study by Bird (1969), but this type of study goes back to 1931 (Heinrich). What this sort of diagram is intended to show is that for every major injury there are increasingly larger numbers of less serious losses. However, accident triangles of this type can be misleading because it is possible to have damage-only accidents which are very serious indeed in financial terms.

Since 1931, there have been various versions of the accident triangle, with different incident categories and different numbers. Several such

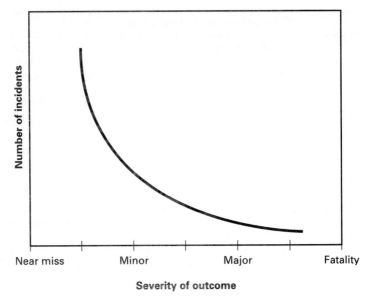

Severity of outcome

Figure 3.1 Relationship between frequency and severity

studies are reported in the HSE publication[2], and these are summarised below:

	'Over 3 days'	*'Minor'*	*'Non-injury'*
Construction	1	56+	3570+
Creamery	1	5	148
Oil platform	1	4	126
Hospital	1	10	195

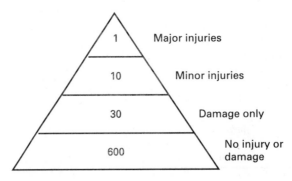

Figure 3.2 Accident triangle

These results demonstrate that the ratios for different outcomes differ from industry to industry and this is likely to be due to the different ranges of risk involved. Note, however, that they continue the conflation of the qualitative and quantitative dimensions.

3.2.3 Practical implications

The preceding discussion has demonstrated that accidents and incidents are subsets of a much wider category of events and that there is no universally agreed definition of the subsets involved.

This means that the first stage in any work on accident and incident data should be a clear definition of the particular subset which will be used. To be of practical value, the definition should deal with the following:

1 The nature of the outcomes to be included, for example injury, sickness or damage to assets and, in particular, whether hazard creation will be considered as an outcome.
2 The severity of the outcomes to be included. This can be difficult since it will usually involve specifying a more or less arbitrary point on a continuum and a decision will have to be made on, for example, near misses.
3 The population to be covered. For example, will it be restricted to the organisation's personnel and assets or will contractors and members of the public be included?

3.3 Collection of accident and incident data

It is assumed for the purposes of this section that the collection of accident and incident data involves three main stages:

1 Ensuring that accidents and incidents are reported. Unless individuals know what accidents and incidents to report, and how to report them, the relevant data are unlikely to be collected.
2 Checking for non-reporting of the types of accident or incident which should be reported and recorded.
3 Recording details of the accidents and incidents which are reported. If the data collected on accidents and incidents are to be of real value, they have to be recorded accurately.

However, before starting a discussion it is necessary to consider some problems with terminology and, in particular, the way 'accident reporting' and 'incident reporting' are used to mean a number of different things.

If we consider the chronology of an accident and its aftermath, we have the following identifiable stages:

1 The person who sustains the injury, or someone else, *reports* that an accident has happened, usually to a supervisor or manager.

2 The person to whom the accident is reported, makes a written record of the salient points, usually on an *accident report form* or *accident record form*.

3 The accident is investigated and, if it is sufficiently serious, is *reported* to the relevant national authority, for example the Health and Safety Executive in the UK.

4 The person who investigates the accident writes a *report* on his or her findings, to which are added any suggestions for remedial action.

5 The person who investigates the accident *reports back* to those involved in the outcome of the investigation and the action to be taken.

Since it is not always obvious from the context which of these uses of 'report' is the relevant one, the following terminology will be used in this chapter:

● *Accident report*. The report made by the person who sustains the injury, or someone else on their behalf.

● *Accident record*. The written record of the salient points, usually on an *accident record* form.

● *Accident notification*. The notification of an accident to the relevant national authority, for example the Health and Safety Executive in the UK.

● *Investigation report*. The written report on the findings of an accident investigation, together with any suggestions for remedial action.

● *Feedback*. The reporting back to those involved on the outcome of the investigation and the action to be taken.

3.3.1 Ensuring accidents and incidents are reported

In general, the less serious an incident is, the less likely it is to be reported. It is extremely difficult in most organisations to 'cover up' a fatality or major injury, but minor injuries often go unreported. However, there are various things which can be done to improve reporting and these are described below.

Have a 'user friendly' system. Reporting and recording systems which are too onerous for the quantity of data to be collected will not be used. For example, using 'major' accident forms for collecting information on 'minor' accidents or incidents will discourage reporting of minor accidents and incidents because the amount of effort required is not perceived as being commensurate with the seriousness of the accident.

Emphasise continuous improvement. The reasons for collecting the data (continuous improvement and prevention of recurrence) should be clearly stated and repeated often.

Avoid a 'blame culture'. If accident or incident reports are followed by disciplinary action or other minor forms of 'blame', people will stop reporting.

Demonstrate that the data are used. If people who have to report and record cannot see that use is being made of their efforts, they will stop making the effort.

Always give feedback. It is not always possible or necessary to take action on a report, but there should be feedback to the people concerned explaining the action being taken or reasons for the lack of action.

The practicalities of implementing these various points will vary from organisation to organisation but any weaknesses in accident and incident reporting systems can usually be identified quite easily in the course of a straightforward review of the systems against the criteria listed above.

3.3.2 Checking for non-reporting

Where it is important to have an accurate measure of the occurrence of a particular category of accident or incident, checks should be made that all of the relevant accidents or incidents are being reported. Three methods of carrying out such checks are described below:

1 Interviews with people who are likely to have experience or knowledge of the relevant accidents or incidents. People are more willing to talk about accidents or incidents they did not report if they are confident that there will be no adverse consequence as a result of their revelations. They are even more willing to talk about accidents or incidents other people did not report, if they know it will not result in adverse consequences for the people being identified. A skilled interviewer who has carried out an appropriate sample of interviews should be able to make a reasonably accurate assessment of the proportion of accidents or incidents which is going unreported.
2 Inspections of locations and people. The simplest example in this category involves inspecting plant and equipment for damage and comparing the findings from the inspection with the most up-to-date damage records. A similar approach can be used for minor injuries by, for example, inspecting people's hands, checking for dressings, cuts, grazes and burns and then comparing the inspection findings with the injury records. However, this approach can engender resentment and should be undertaken with care. In some organisations, the dressing for minor injuries are a characteristic colour, easily recognised even from a distance. The use of this type of dressing makes inspections for minor injuries much easier.
3 Cross-checking one set of records with another. The usefulness of this type of technique will depend on the records available and their accuracy, but possible cross-checking includes the following:

   ● Where there are records of what is taken from a first aid box, these can be checked against injury records to see whether everyone who has made use of the first aid box has reported an injury.

- Where there are records of plant and equipment maintenance, these can be checked against records of accidental damage to plant and equipment to see whether all of the relevant repairs which have had to be carried out have been recorded as accidents.
- Where records of 'cradle to grave' or 'mass transfer' are available for particular chemicals or substances, these can be used to check whether unexplained losses of chemicals or substances have appeared as, for example, accidental spillage records.

Any, or all, of the above techniques can be used to check the adequacy of reporting and to ensure 'good' data are available for analysis.

3.3.3 Recording details of accidents and incidents

Details of accidents and incidents are usually recorded on some type of form and the design of this form can have a marked influence on what gets reported. When designing for recording accident data the following points should be taken into account:

- The form should require only those data which it is reasonable to expect people will be willing to record. Many forms are designed to cover all of the eventualities of a major injury with boxes for whether the accident has been notified, next of kin, and a range of other details. People are then expected to use this form to record details of a cut finger! It could be argued that there is no need for a form for serious accidents since they will all be notified on an official form (see below), investigated in detail and an accident report written. Even if such a form is considered necessary, it should be used only for the purpose for which it was designed, and a separate form designed to use for the recording of less serious accidents and incidents.
- The form should require only those data which it is reasonable to expect people to be competent to provide. Many forms include spaces for such things as 'root cause of accident' and 'suggestions for risk control measures' and expect them to be filled in by people without the competences to provide accurate data. In the worst cases, these data are then analysed as though they had the same validity as the data it is reasonable to expect will be accurate, such as time of incident and part of body injured. There are two solutions to this, either omit from the form items requiring judgement, or provide the necessary competences.
- Ensuring completion of the form should be the responsibility of people at an appropriate level in the organisation. It is reasonable to expect work people to report minor injuries, near misses and hazards but it is not necessarily the case that these people are willing or able to record the details necessary for effective analysis.

There are various ways of meeting the requirements listed above but they all depend on well-designed forms and well-thought out systems for reporting and recording. A suitably designed accident report form may also be accepted by insurers as notification of a claim under their EL insurance policy.

3.4 UK legal requirements to notify accidents and incidents

Accident notification requirements are specific to a particular country and readers outside the UK should identify the requirements of their local legislation, or requirements imposed through other means. The remainder of this section summarises the UK requirements for accident notification. However, this is only a summary and detailed study of the legislation referred to is essential for safety professionals and those responsible for notification.

3.4.1 The Reporting of Injuries, Diseases and Dangerous Occurrences Regulations 1995

The Reporting of Injuries, Diseases and Dangerous Occurrences Regulations 1995 (RIDDOR) with its supporting guide[3] place duties on employers and the self-employed to report certain incidents which occur in the course of work. These reports are used by the enforcing authorities to identify trends in incident occurrence on a national basis. The reports also bring to the attention of the enforcing authorities serious incidents which they may wish to investigate. Reports must be made by the 'responsible person' who, depending on circumstances, may be an employer, a self-employed person, or the person in control of the premises where the work was being carried out.

The methods of reporting depend on the type of incident. For an incident resulting in any of outcomes listed in *Table 3.2* the relevant enforcing authority must be notified by the quickest practicable means, usually by telephone.

This notification must be followed by a written report within 10 days using Form F2508, details of which are given below. If they wish, the enforcing authorities can make a request for further information on any incident.

Dangerous occurrences are, in general, specific to particular types of machinery, equipment, occupations or processes and knowledge of the relevant incidents is necessary to ensure proper reporting. Some examples are given in the second part of *Table 3.2* illustrating the range of incidents involved.

An accident, other than one causing a major injury, which results in a person 'being incapacitated for work of a kind which he might reasonably be expected to do . . . for more than three consecutive days (excluding the day of the accident, but including days which would not have been working days)' is referred to as a 'three day' accident and is only required to be notified by a written report.

Table 3.2. Incidents to be reported by the quickest practicable means

Fatalities

Major injuries as listed below:

Fractures (other than finger, thumb or toe).

Amputations.

Dislocations of shoulder, hip, knee or spine.

Loss of sight (temporary or permanent).

Chemical or hot metal burn to the eye or any penetrating injury to the eye.

Electric shock or burn leading to unconsciousness, or requiring resuscitation, or admittance to hospital for more than 24 hours.

Any injury leading to hypothermia, heat induced illness or to unconsciousness, or requiring resuscitation, or requiring admittance to hospital for more than 24 hours.

Loss of consciousness caused by asphyxia or by exposure to a harmful substance or biological agent.

Either of the following conditions which result from the absorption of any substance by inhalation, ingestion or through the skin: (a) acute illness requiring medical treatment; (b) loss of consciousness.

Acute illness which requires medical treatment where there is reason to believe that this resulted from exposure to a biological agent or its toxins or infected material.

Dangerous occurrences. Specified incidents involving:

Lifting machinery (includes fork-lift trucks) – collapse, overturning, or failure of any load bearing part.

Pressure systems – failure of any closed vessel or associated pipework, where the failure has the potential to cause death.

Freight containers – failure of container or load bearing parts while it is being raised, lowered or suspended.

Overhead electric lines

Electrical short circuit leading to fire or explosion resulting in plant stoppage for more than 24 hours, or with the potential to cause death.

Explosives

Biological agents

Malfunction of radiation generators

Breathing apparatus

Diving operations

Collapse of scaffolding

Train collisions

Wells (NB not water wells)

Pipelines

Fairground equipment

Carrying of dangerous substances by road

Collapse of building or structure

Explosion or fire which results in stoppage or suspension of normal work for more than 24 hours where the explosion or fire was due to the ignition of any material.

Escape of flammable substance the sudden uncontrolled release, inside a building, of e.g., 100 kg or more of a flammable liquid or 10 kg or more of a flammable gas. If in the open air, 500 kg or more of flammable liquid or gas.

Escape of substances in any quantity sufficient to cause the death, major injury, or any other damage to the health of any person

Fatalities, major injuries, dangerous occurrences and three day accidents have to be reported on Form F2508. The main requirements for information on this form are:

- Date and time of the accident or dangerous occurrence.
- For a person injured at work, full name, occupation and nature of injury.
- For a person not at work, full name, status (e.g. visitor, passenger) and nature of injury.
- Place where incident happened, brief description of the circumstances, date of first reporting to the relevant authority and method of reporting.

If an employee dies within one year as a result of an accident the employer has to inform the enforcing authority as soon as he learns of the death, whether or not the accident had been reported originally.

If a person at work suffers from an occupational disease and his or her work involves one of a specified list of substances and activities, the responsible person must send a report to the relevant enforcing authority as soon as he learns of the disease. *Table 3.3* gives some examples of the diseases and associated activities listed in Schedule 3 of RIDDOR. Notification of occupational diseases is normally by Form F2508A, the main requirements of which are:

- Date of diagnosis of disease.
- Name and occupation of person affected.
- Name and nature of disease.
- Date first reported to the relevant authority, and method of reporting.

Copies of F2508 and F2508A are contained in the guide[3].

- Records of reportable incidents must be retained for at least three years. This can be as photocopies of the Forms F2508 and F2508A or the data can be kept on computer when registration under the Data Protection Act will be necessary.

RIDDOR covers:

- employees
- the self-employed
- those receiving training for work
- members of the public, pupils and students, and other people who suffer injuries or diseases as a result of work activities

but does not cover:

- patients who die or are injured while undergoing medical or dental treatment
- some incidents on board merchant ships

Table 3.3. Some example of occupational diseases and associated activities

Diseases	Activities
Cataract due to electromagnetic radiation	Work involving exposure to electromagnetic radiation (including heat).
Cramp of the hand or forearm due to repetitive movements	Work involving prolonged periods of handwriting, typing or other repetitive movements of the fingers, hand or arm.
Beat hand, beat elbow and beat knee	Physically demanding work causing severe or prolonged friction or pressure on the hand, or at or about the elbow or knee.
Hand arm vibration syndrome	Work involving a specified range of tools or activities creating vibration.
Hepatitis	Work involving contact with human blood or blood products, or any source of viral hepatitis.
Legionellosis	Work on or near cooling systems which are located in the workplace and use water, or work on hot water service systems located in the workplace which are likely to be a source of contamination.
Rabies	Work involving handling or contact with infected animals.
Tetanus	Work involving soil likely to be contaminated by animals.
Tuberculosis	Work with persons, animals, human or animal remains or any other material which might be a source of infection.
Poisoning by specified substances including mercury and oxides of nitrogen	Any activity.
Various cancers	Various activities.
Occupational dermatitis	Work involving exposure to a range of substances.
Occupational asthma	Work involving exposure to a range of agents.

- death or injury where the Explosives Act applies
- death or injury as a result of escapes of radioactive gas
- cases of agricultural poisoning
- where the incident is reportable under the Road Traffic Act.

3.4.2 Other legislative requirements

Serious incidents as defined have to be recorded and reported to the relevant authority. However, there is also a requirement to record details of less serious incidents, for example minor injuries, although these do not have to be reported to an authority. This requirement is imposed by the Social Security (Claims and Payments) Regulations 1979 but it does not apply to all employers. However, where it does apply, these less serious incidents can be recorded either in an Accident Book (BI 510) or on an organisation's own form or forms. In either case, the minimum information which must be recorded is:

- Full name, address and occupation of injured person.
- Date and time of accident.
- Place where accident happened.
- Cause and nature of injury.
- Name, address and occupation of person giving the notice, if other than the person injured.

Where an organisation is using its own form, additional information for internal use can be recorded. As with the serious incidents, if minor injury records are kept on computer, registration under the Data Protection Act will be required.

3.5 The use of accident and incident data
3.5.1 Introduction

In this section a more detailed look is taken at how accident and incident data can be used to learn from what has gone wrong in the past so that risk control measures can be implemented or improved. There are three main aspects:

1 Measuring whether performance is improving or deteriorating using trend analysis.
2 Making comparisons using accident and incident data.
3 Learning from accident and incident occurrence by using epidemiological analysis.

3.5.2 Trend analysis

By making continuous measurements of the numbers of accidents and incidents, it is possible to make comparisons of performance in different time periods and compare one with another, that is, carry out trend analysis over time. However, any such analysis can be influenced by changes other than changes in the effectiveness of the safety management. For example, if an organisation is reducing the amount of work it does it is likely that the number of accidents will decrease, whether or not there are any changes in safety management practices. This is an obvious, but important, point. If a press operates one million times, or delivery drivers drive one million miles, there is a certain scope for accidents. If a reduction in work halves the number of press operations, or the number of miles driven, the scope for accidents is reduced. Similarly, if the amount of work being done is increasing, we would expect the number of accidents to increase.

Because numbers of accidents can be influenced by these sorts of changes, the trend analysis will be dealt with in two stages. First, assuming that there is a steady state, with no relevant changes, this will allow study of the basic techniques without undue complication. Second, the techniques required to take into account the sorts of changes described above will be considered.

3.5.2.1 Trend analysis with a steady state

The most straightforward method of trend analysis is to plot the numbers of accidents or incidents against a suitable measure of time. However, it is also possible to plot on a graph not just the numbers of accidents and incidents but also some measure of severity. Typical examples of this sort of measure include days lost through sickness, litres of lost fuel, cost of damage repair, etc. Typical time measures include monthly, quarterly and annually. *Figures 3.3 to 3.6* show examples of these sorts of plots.

One practical problem with graphs is that the more detailed they are, the more difficult it is to judge the trend 'by eye'.

Compare the two graphs shown in *Figures 3.7 and 3.8*.

Although using the same data for both *Figures 3.7 and 3.8*, it is easier to see from the quarterly plot that there appears to be a slight downward trend. Generally grouping data in this way 'smooths out' variations and makes trends easier to identify. One technique for smoothing data, the quarterly moving mean, is quite simple and the steps required are described below, using as an example the data from *Figure 3.7*.

- For the first two months of the year, the accident numbers are plotted on a month-by-month basis as in *Figure 3.7*.
- For the third month, the numbers of accidents for January, February and March are added together and the result divided by three to give the mean number of accidents and this is plotted[1]. While this

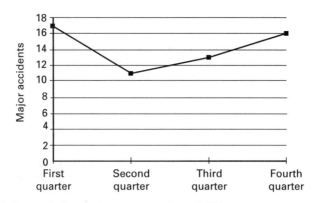

Figure 3.3 Quarterly Figures for major accidents (1998)

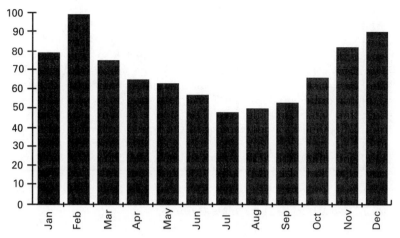

Figure 3.4 Days lost per month through sickness (1998)

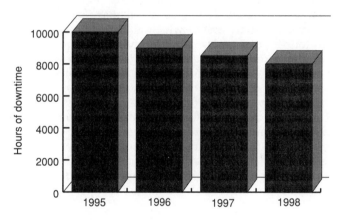

Figure 3.5 Hours downtime by year (1995 to 1998)

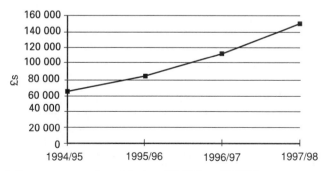

Figure 3.6 Damage costs – financial years 1994/95 to 1997/98

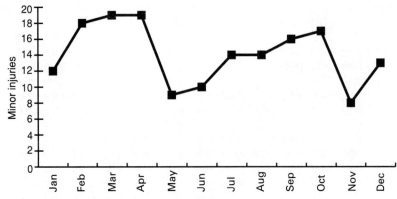

Figure 3.7 Monthly minor injury figures (1998)

calculation may be more familiar as giving the 'average' number of accidents per month, it should be noted that there are a number of different averages, only one of which is the mean.

● For the fourth, and subsequent months, the quarterly mean is calculated from the current month's plus the previous two months' figures divided by three and the results of these calculations plotted.

The quarterly moving mean for the data in *Figure 3.7* is illustrated in *Figure 3.9*. Note that it is usual to plot both the actual monthly figures as well as the moving mean, and this has been done in *Figure 3.9*.

Whichever method of trend analysis is used, a check should be made that any change in direction is more than a random fluctuation.

Suppose that in a particular year there were 100 accidents in a company and that in the following year the company proposed to carry out the

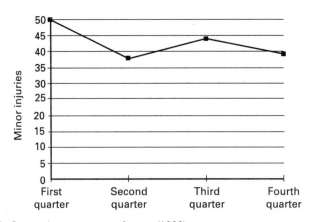

Figure 3.8 Quarterly minor injury figures (1998)

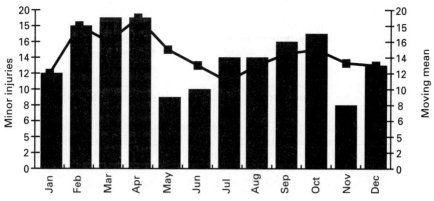

Figure 3.9 Monthly accidents and quarterly moving mean (1998)

same amount of work with no changes which would affect risk. In these circumstances, we would expect around 100 accidents in the year following the one for which records were available. Note that we would not expect exactly 100 accidents, but *around* 100 accidents. If there were 99 accidents or 101 accidents we would be able to say that this was due to random fluctuation and, more generally, anything between say 95 and 105 accidents could also be random.

The difficulty arises when the number of accidents reaches 85 or 90. Are these numbers due to random fluctuation, or is someone doing something which is improving risk control and influencing the accident numbers? Statisticians refer to fluctuations in numbers which cannot reasonably be attributed to random fluctuation as 'significant' when they may make statements like: 'There is only a 5% chance that the improvement in accident performance is due to random fluctuations', or 'This deterioration in accident performance would have occurred by chance in only 1% of cases'.

The working out of the significance of fluctuations in numbers has practical importance in the more advanced techniques of loss management since we can only draw valid conclusions when we know whether or not particular fluctuations in numbers are significant. For this reason, it is valuable to have some idea of the significance of fluctuations and trends. One way of doing this is to use historical accident data and work out upper and lower limit lines, based on the mean of these data. If we used this technique on the data shown in *Figure 3.9*, we could draw up a chart for 1999 which would look like the one shown in *Figure 3.10*.

As the monthly accident figures for 1999 become available, they are plotted on the chart in the usual way. Monthly numbers of accidents which are within the limit lines are defined as random fluctuations. Only if the number of accidents is above the upper limit line, or below the lower limit line, is the fluctuation considered significant.

Using this type of upper and lower limit line has practical advantages since it can prevent resources being expended on attempts to reduce

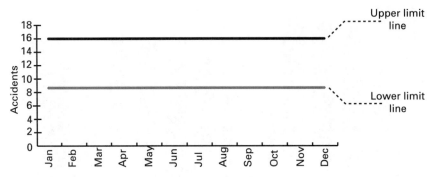

Figure 3.10 Illustration of upper and lower limit lines

increases in accidents which are purely random. While it might be argued that no resources spent on attempts to reduce accidents are wasted, resources are always limited and it is preferable to use them where there is good statistical evidence that they will do the most good.

The details of the calculations required for upper and lower limit lines, and related statistical techniques such as confidence limits and control charts, are beyond the scope of the present chapter, but details can be found in references 4 to 7 inclusive.

3.5.2.2 Trend analysis with variable conditions

So far, for the sake of simplicity, we have assumed that everything has remained stable in the organisation. In the real world, however, things rarely remain the same for any length of time and we need methods of trend analysis which can take this into account.

In an ideal world, we would be able to measure changes in risk in an organisation and hence determine how well the risk was being managed. For example, the measures would enable us to say such things as 'despite a 50% increase in risk due to additional work being done, the accidents increased by only 25%', or 'there was a 10% reduction in risk because of the new machines and work procedures, but accidents increased by 5%'.

Unfortunately, it is rarely possible to measure risk in this sort of way so what we have to do in practice is to find some proxy for risk which we can measure and use instead. Two such proxy measures in common use are numbers of people employed and numbers of hours worked which are used to calculate two accident rates.

● *Incidence rate.* This index gives the number of accidents for 1000 employees and is used to take into account variations in the size of the workforce:

$$\text{Incidence rate} = \frac{\text{Number of accidents} \times 1000}{\text{Number employed}}$$

- *Frequency rate*. This index gives the number of accidents for every 100 000 hours worked and takes into account variations in the amount of work done, and allows for part-time employees.

$$\text{Frequency rate} = \frac{\text{Number of accidents} \times 100\,000}{\text{Number of hours worked}}$$

There are, however, a number of problems with these accident rates.

- *Terminology.* Although the versions given above are in general use, there is no universal agreement as to the basic formula. A rate cannot be interpreted unless the equation on which it was based is known.
- *Definitions.* There is no general agreement on what constitutes an 'accident', with some organisations basing their rates on only major injuries, while others use both major and minor injuries. Similarly, there is no general agreement on what constitutes an employee – incidence rates can be reduced by employing more part-time people! Hours present similar problems, with different types of hours having significantly different types of risk. For example, 'working' time, when the risk is high, versus 'waiting' time, when the risk is low. Also, 'staff' do not normally book their time although they can face the same risks as hourly paid employees.
- *Multipliers.* There is no general agreement on which multipliers should be used and it is normal to select one that suits the particular organisation.

In general, the value of accident rate figures depends on the quality of the data on which they are based and the honesty of the person preparing them. Quoted rates should always be treated with caution until the basis of the calculation has been determined.

Using *incidence* and *frequency rates* enable sensible trend analyses to be carried out during periods when changes are being made in the organisation that affect the number employed or the amount of work being done and, used properly, they can provide useful safety information.

These rates also enable us to make comparisons between one organisation and another, or between different parts of the same organisation, i.e. enable comparative analyses to be made.

3.5.2.3 Comparisons of accident data

It is only possible to make valid comparisons when there is some measure of the risk being managed. When the numbers employed or the hours worked are taken into account, these are only a proxy for risk and are used because we can measure them, rather than because they are good indicators of risk.

Considering two organisations, each with a frequency rate of 100, this could be because:

- The organisations have roughly equal levels of risk and are managing them equally effectively.
- One organisation has high levels of risk and is managing them well, while the other organisation has low levels of risk and is managing them badly.

This should be borne in mind when making, or interpreting, comparisons of accident data since it is a fundamental weakness of such comparisons. In general, a comparison will be valid only to the extent that the risk levels in the organisations being compared are equal.

Having dealt with this caveat, the types of comparison which can, with reason, be made are:

- *Comparisons between parts of the same organisation.* In theory, these are the simplest and potentially most accurate comparisons. This is because the measurement of risk, the definition of what has to be reported, reporting procedures, and methods of calculation are all under the organisation's control and can be standardised. However, the value of this comparison depends on the effectiveness of the reporting, which may not be consistent throughout the organisation.
- *Comparisons between one organisation and another.* Industries in the same sector can compare accident data one with another, assuming that they are willing to do so. In the UK, for example, there are national associations for particular industry and service sectors which provide a forum for comparing accident data. More formally, there have been moves recently to include accident data in 'benchmarking' exercises where organisations compare various aspects of their performance with those of their competitors.
- *Comparisons between an organisation and the relevant industry or service sector.* Some trade organisations publish aggregated data on accidents for their industry or service sector giving, for example, the 'average' frequency and incidence rates for a particular year. Examples of these sorts of accident data for the UK are published by the HSE annually[8]. The HSE's Epidemiology and Medical Statistics Unit also produces statistics on occupational ill health.
- *Comparisons between countries.* Where appropriate data are available, comparisons can be made between accidents in one country and another, either for the country as a whole, or by industry or service sector. However, there are major variations in accident reporting procedures between countries so that comparisons of this type should be made with great care.

A particular problem with all of these comparisons is that there is no consistency about what constitutes an 'accident' and it should be remembered that this was one of the problems with any comparison of incidence and frequency rates. One way of improving comparisons is to calculate a rate which takes into account the severity of the accidents, i.e. the number of days lost per accident, to give the mean duration rate:

$$\text{Mean duration rate} = \frac{\text{Number of days lost as a result of } x \text{ accidents}}{x \text{ accidents}}$$

This mean duration rate can be used in trend analysis in the same way as other rates. A disadvantage of it is that it can give a misleading picture since it can show a decrease when the numbers of days lost is increasing, i.e. more accidents but fewer days lost per accident.

For this reason, some organisations use an alternative severity rate:

$$\text{Severity rate} = \frac{\text{Number of days lost as a result of accidents}}{\text{Number of hours worked}}$$

The final point to make on comparisons is that the rates described above should, when the relevant data are available, be used in conjunction with each other. This is because they do not necessarily give the same result, as is illustrated, using simplified data, in *Table 3.4*.

3.5.2.4 Accidents and incidents as a measure of risk

Accurate accident and incident data will provide a measure of what has gone wrong in the past, and allow comparisons over time (trend analyses) and comparisons between one organisation and another. What these data will not do, even if they are accurate, is to provide a measure of risk.

Information on the number of accidents gives us very little information about risk. Two organisations can have the same number of accidents because one is managing high levels of risk very well, while the other is managing low levels of risk very badly. Alternatively, because risk is probabilistic, two organisations with the same levels of risk can have widely different numbers of accidents because one was 'lucky' and the other was not.

True levels of risk in an organisation can only be determined accurately using appropriate risk assessment methodologies, details of which will be found in Chapter 8 of *Workplace and Safety*.

Table 3.4. Comparisons using incidence, frequency and severity rates

	A	B	C	D
Number of accidents	100	80	60	20
Numbers employed	100	40	60	20
Incident rate	1000	2000	1000	1000
Hours worked	10 000	8000	3000	2000
Frequency rate	1000	1000	2000	1000
Days lost	100	80	60	40
Mean duration rate	1	1	1	2
Severity rate	0.01	0.01	0.02	0.02

3.6 Epidemiological analysis

3.6.1 Introduction

The techniques of epidemiological analysis were first applied to the study of disease epidemics and historical example will be looked at by way of illustration to show how epidemiological techniques can be applied to accident and incident data.

Typhoid plague was a major cause of death in cities for many years. No one knew what caused the plague but many doctors looked for patterns in where the epidemics occurred. This was done on a trial and error basis with different people looking at where plague victims lived, what they ate, and the work they did. Eventually it was discovered that plague epidemics were centred around certain wells from which the city dwellers of those days obtained their drinking water. It was also found that closing these wells stopped the spread of the plague in those areas. Although no one knew why the wells, or the water from them, was causing the plague, they had found an effective way of stopping the plague spreading. In fact, it was many years before the water-borne organisms responsible for plague infection were identified.

This example illustrates the essential elements of epidemiological analysis. It is the identification, usually by trial and error, of patterns in the occurrence of a problem which is being investigated. These patterns can then be analysed to see whether causal factors can be identified and remedial action taken.

Epidemiology is used to identify problems which would not be apparent from single incidents. For example, if accidents occurred more frequently at a particular type of location, the records provide a guide to where investigation will be most fruitful and cost effective, although they provide no information on the possible causes.

3.6.2 Techniques of epidemiological analysis

Epidemiological analysis is only possible when the same type of information (data dimension) is available for all (or a substantial portion) of the accidents being analysed. Typical data dimensions include location and time of the accident or incident, the part of the body injured in an accident and the nature of the injury.

The simplest form of epidemiological analysis is *single dimension analysis*. This involves comparing incidents in the population on a single data dimension, for example time of occurrence or nature of injury. The analyst would look for any deviation from what would reasonably be expected. For example, if work is spread evenly over the working day, we would expect times of injuries also to be spread evenly. Where peaks and troughs are found in accident occurrences, these should be investigated. The analysis is slightly more complicated when an even spread is not expected as the analyst has to carry out preliminary work to determine the expected spread.

The analyst will look for both over-representation and under-representation when carrying out the analysis. Both should be investigated, over-representation because it suggests that there are risks which are being managed poorly, under-representation since it suggests either a degradation in the reporting and recording system, or particularly effective management of risk from which others might learn.

The principles and practices described above for single dimension analysis can also be applied to two or more dimensions analysed simultaneously, this is referred to as *multi-dimensional analysis*. This type of analysis can identify patterns which would not be apparent from analysing the data dimensions separately and examples include part of body injured analysed with department, and time of day analysed with nature of injury.

Full-scale epidemiological analysis of a set of data will involve analysis of all of the single data dimensions separately and analysis of all of the possible combinations of these single dimensions. For this reason, epidemiological analysis is a very time consuming process and where more than a trivial number of data are involved, the only practical approach is to use a computer. Suitable software for epidemiological analysis is described later in the chapter.

The epidemiological analysis merely identifies patterns in data distribution, it does not give information on why these patterns are occurring. This can only be determined by appropriate follow-up investigations and these are dealt with in the section on accident investigation.

3.6.3 Epidemiological analysis with limited data

The fact that the detailed data described earlier as necessary for full-scale epidemiological analysis does not prevent the techniques being applied to information that had already been gained.

Valuable results can often be obtained simply by tabulating accident data for the past two or three years and looking for patterns in accident occurrence. It is also worth trying to discover if there were no accidents for particular places, times, people, etc. since this can provide clues on non-reporting or effective risk control measures.

3.7 Accident investigation

3.7.1 Introduction

Accident investigations can be carried out for a number of reasons, including:

- Collecting the information required for reporting the accident to the enforcing authorities.
- Establishing where the fault lay.

- Obtaining the information required to pursue, or defend, a claim for damages.
- Obtaining the information necessary to prevent a recurrence.

In theory, a thorough investigation will result in the collection of the information required to satisfy all of these purposes but, in practice, this is rarely the case. If, for example, the primary purpose is to collect the information required for accident notification then the investigation is usually stopped when the relevant information has been collected, whether or not this information includes that required for the prevention of a recurrence. When the primary purpose is to establish where the fault lay, if this is allowed to extend to who was responsible, there may be an additional problem in that the investigation may become adversarial, that is, the investigators are on one 'side' or the other, for example the employer's 'side' or the injured person's 'side'. This can lead to biases in data collection with, for example, information which does not support a particular investigator's 'side' being ignored or not recorded.

The ideal investigation is, therefore, one which is neutral with respect to fault and has the primary purpose of obtaining the information necessary to prevent a recurrence.

In all accident investigations of this type there are two types of information to collect:

- Information about *what* happened which is usually factual and has limited scope for interpretation, for example the date and time of the incident, and what caused the injury, damage or other loss.
- Information about *why* it happened is concerned with the causes of the incident. It is more difficult to identify and more open to interpretation.

This distinction between 'what' and 'why' corresponds with the terminology used elsewhere to make roughly the same distinction. Typical terms include:

- *Immediate or proximate* causes are the direct causes of the injury, damage or other loss.
- *Underlying or root* causes are the reasons why the accident or incident happened.

These terms are used throughout the remainder of this chapter.

Collecting information about what happened is the essential first step in an investigation and must be completed before considering why it happened.

3.7.2 Collecting information on what happened

The two main sources of information are observation of the accident site and interviews with those involved (the injured person, witnesses, those who rendered assistance and so on). Observation of the site is fairly straightforward but interviewing is a skill which has to be learned. There are a number of key points to be followed for good interviewing.

3.7.2.1 Interviewing for accident investigations

There are three important aspects of interviewing which have to be considered:

- Coverage
- Keeping an open mind
- Getting people to talk.

(a) *Coverage*
This aspect of interviewing deals with the nature and amount of information which has to be collected, how to decide when all the relevant information has been obtained and how to avoid collecting information which is of no value?

What is relevant and valuable will, of course, depend on the purpose of the investigation and as a general guide, coverage should include all the information necessary to enable a decision to be made about the appropriate remedial action. However, in this first stage of the investigation, the purpose is to establish a clear idea of what happened. The information required falls into two categories:

1 Information which is common to all types of incident and which is best dealt with by using a pro-forma containing spaces for the information required. The accident record form used for this purpose should include information which gives:

   - Details of the incident – e.g. time, date and location.
   - Details of person injured – e.g. names, age, sex, occupation and experience.
   - Details of the injury – e.g. part of body injured, nature of injury (cut, burn, break etc.), the agent of injury (knife, fall, electricity etc.), and time lost.
   - Details of asset or environmental damage – e.g. what was damaged, nature of damage, and the agent of damage.

   It is this type of information which is best used for the sorts of analyses discussed earlier since it is common to all incidents and can, therefore, be used for trend, comparison and epidemiological analyses.
2 Other information has to be recorded as a narrative and space for this should be included on the accident record form. However, it is often necessary for this brief summary to be supplemented by a more detailed investigation report.

(b) *Keeping an open mind*
One of the main difficulties during an investigation is avoiding assumptions about what has happened. The greater the experience of the

type of site involved, the nature of the work and the people, the more likely is it that assumptions will be made. There is always the possibility that an investigation will result in a summary of what was thought likely to have happened, rather than what actually happened.

To avoid making assumptions questions should be asked about all aspects of what happened, even if the answer is known. Perhaps even especially when confident of what the answer will be!

Making assumptions can lead to forming an inaccurate picture of what happened, which in turn can have serious implications if it leads to suggestions for remedial actions which are wholly inappropriate. Where possible remedial action is identified early in the investigation, this is a warning sign that too many assumptions may have been made.

(c) Getting people to talk

Interviewees will volunteer information more readily if a rapport can be established and maintained with them. Rapport is the term used to describe the relationship between people which enables a ready flow of conversation without nervousness or distrust. A wider range and more accurate information can be collected when a rapport has been established with the people being interviewed. There are no techniques which will guarantee that rapport is established, but the guidelines listed below will make it more likely:

(i)   *Interview only one person at a time.* It is difficult to establish rapport with two or more people simultaneously since each will require different responses. This may not be possible in some circumstances, for example if the person interviewed requests that a representative attends. In these circumstances, the status of any attendees should be clearly established at the start of the interview including whether they are just observers, will be answering questions on the interviewee's behalf or whether they will be entitled to interrupt.

(ii)   *Have only one interviewer at a time.* 'Board' or 'panel' interviews should be avoided since they require the interviewee to communicate with more than one person, and this is rarely successful. Note, however, that there are many circumstances where it may be necessary for more than one person to be involved in the interview. For example, the employee's representative may wish to be involved. In these circumstances, the interviewer should lead the interview and invite the second interviewer or representative to ask questions at an appropriate point. This procedure should be explained to the interviewee and his representative at the start of the interview so that it has a minimal effect on rapport.

(iii) *Introduce yourself and explain the purpose of the interview.* Do this even if you have already been introduced by someone else. The interviewee will gain confidence if he or she knows who you are and why the interview is taking place. Emphasise that the primary purpose of the interview is the prevention of a recurrence and that action will be taken on the results of the investigation.

(iv) *Check the interviewee's name and the part they played in the incident.* This may sound obvious but checking before the interview can save embarrassment later on. Confusion can arise when, for example, more than one person has been injured, where more than one accident has occurred in the same area, or where other interviews are in progress for some different purpose, for example work study.

(v) *Start the interview on the interviewee's home ground.* The idea is to start the interview with things which are familiar to the interviewee and hence establish a rapport, then move on to the details of the accident. This is helped by beginning the interview at the interviewee's place of work and talking about their normal job before moving on to discussion of the accident.

It is important to establish rapport before moving on to collect detailed information. If this is not done, the interview may degenerate into a series of stilted questions and one word answers. This can also happen if rapport is not maintained and there are a number of things which will help maintain rapport:

(vi) *Prevent interruptions.* Make sure the interview is not interrupted. Interruptions come from other people and an effective way of preventing this is by choosing a suitable place for the interview where interruptions are unlikely. However, the interviewer can interrupt the interview by stopping the interviewee to ask questions. In general it is best to let the interviewee talk and ask any questions when he or she gets to a natural break in their story.

(vii) *Use open questions rather than closed questions.* Open questions are ones which cannot be answered with 'yes' or 'no'; closed questions are ones which can be answered with a 'yes' or 'no'. For example, 'What was the noise level like?' is an open question, 'Was it noisy?' is a closed question. In general, closed questions should be used only to check on specific points already made by the interviewee.

(viii) *Avoid multiple questions.* For example, a question such as 'Can you tell me what everyone was doing at the time?' is better asked as a series of questions starting with 'Can you tell me who was there at the time?' and then a single question about what each of them was doing. Asking multiple questions is likely to result in only part of the question being answered.

(ix) *Keep your manner positive and uncritical.* Interviewees will form an opinion of your manner based on what you say and on your body language. Avoid expressing your views and opinions during the interview, especially if these are critical of what the interviewee has done or not done. Similarly, avoid such obvious signs of lack of interest as not listening, yawning or looking at your watch.

3.7.2.2 Recording the interview

It is essential that written notes are taken during an interview for a number of reasons:

(a) *So that what has been said is not forgotten.* Most people believe that their memory is much better than it really is. Few people can remember all the relevant facts raised during even a short interview.
(b) *So that there can be no confusion over what different people have said.* In most investigations more than one person will have to be interviewed and unless notes are made of each interview it is unlikely that who said what will be remembered, especially if there is a delay between the interviews and writing the report.
(c) *So that the interviewee's narrative is not interrupted.* The importance of not interrupting was mentioned earlier. It is a help in avoiding this if questions are written as they occur ready to be asked at a later and more suitable time. This means that the interview is not interrupted and the points to be raised are not forgotten.

Making notes during the interview is difficult at first but it is a skill, and like all skills can be learned with practice. This skill should be practised whenever possible, and the following should be borne in mind:

(i)   *Timing.* Wait until rapport has been established before starting to take notes. Establishing rapport is difficult enough without the added distraction of note taking.
(ii) *Agreement.* Always tell the interviewee that notes will be taken and get their agreement to this.
(iii) *Content.* Make notes of everything that is said. Even parts of what the interviewee says that seem irrelevant should be recorded. Their relevance should be judged when all the information has been collected, from this and other interviews.
(iv) *Take your time.* Note taking shows the interviewees that what they are saying is of interest. They do not consider it an interruption and are usually happy to wait while notes are made.
(v)   *Review.* At the end of the interview go over the notes with the interviewee checking that what has been written down is an accurate record of what has been said.

3.7.3 Collecting information on why things happen

Once what has happened in an accident has been clearly established, the reason why it happened (the causes) can be investigated. There are various approaches to ensuring adequate coverage of possible accident causes and three options are described below:

1 One or more of the models of human error, such as the one devised by Hale and Hale[9], are summaries of the ways in which human beings think and act and, in particular, how failures in thinking and acting can result in errors. Familiarity with models of this type will help structure an interviewer's approach to the human error aspects of the accident or incident.

2 The Domino Theory provides a succinct description of how the organisational aspects of accident and incident causes link with individual losses, and how human errors can be the result of organisational arrangements. Familiarity with this theory and its variants will help the interviewer avoid too narrow a concentration on the role of the injured person to the exclusion of broader organisational issues.
3 The approach described in an HSE publication[10] is particularly useful for those organisations which have adopted the HSE's Safety Management System since it facilitates the identification of accident and incident causes in terms of weaknesses in the existing Safety Management System.

The next section describes, in outline, how one of these approaches, the Domino Theory, can be used as a means of identifying more accurately the required remedial actions.

3.7.3.1 The Domino Theory

There are various versions of the Domino Theory and the one illustrated in *Figure 3.11* is a generalised version. The basic idea behind the Domino Theory is that individual errors take place in the context of organisations and a useful concept for illustrating them is a series of dominoes standing on end.

If one of the dominoes to the left of the Loss domino falls, it will knock over those to the right and a loss will occur. For example:

- Lack of supervision (management control) results in a situation where oil can be spilt and not cleared up.
- An unsafe act occurs, spilling oil and not clearing it up.

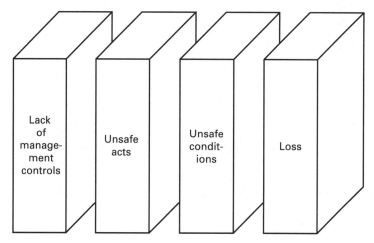

Figure 3.11 Generalised Domino Theory

- An unsafe condition results, a pool of oil on the floor.
- A loss occurs when someone slips on the oil, falls and breaks an arm.

When we investigate the loss, we can identify unsafe conditions, unsafe acts and lack of management controls and establish causes for these, as well as causes for the loss. Continuing the example:

- Possible causes of a person slipping on a patch of oil might be not looking where they were going, or not wearing appropriate footwear.
- Possible causes of not clearing up spilled oil might be lack of time, or not seeing it as part of the job.
- Possible causes of spilling oil might be working in a hurry, inappropriate implements or a poor method of work.
- Possible causes of poor management control might be excessive pressure for production (resulting in hurrying), lack of funding for proper implements, or insufficient attention to designing appropriate systems of work.

The further the cause of the accident is to the left of the dominoes, the greater the implications for lack of management control. By inference it follows that lack of appropriate systems of work may apply to a large number of operations, not just to those which can result in oil spillages. Thus it may be possible to identify and remedy failures in management controls and hence the potential to eliminate large numbers of losses. The usefulness of the investigation can, therefore, extend beyond simply preventing that accident happening again.

One way of doing this is to look systematically at what each of the dominoes represents, determine which one started the fall and concentrate investigation in that area. But it is important to remember that there is rarely a single function or action that causes a particular domino to fall, rather there are a number of reasons which contribute to the fall. There is a need to continue to ask why things happened until all of these contributory causes have been identified. The oil spillage example used earlier illustrates this.

Possible reasons for the person slipping on the oil were that he was not looking where he was going and that he was wearing inappropriate footwear. The question 'Why' should be asked about each of these to see whether further useful information can be obtained. For example, not wearing appropriate footwear could be because:

- he did not know he should be wearing special footwear
- he did not know which type of footwear was appropriate
- the appropriate footwear was uncomfortable
- the appropriate footwear was too expensive
- and so on.

The different answers to these 'why?' questions will have different implications for remedial action so it is important to establish the reason 'why' before making any recommendations. A similar technique should be applied to the other dominoes and again this can be illustrated using the oil spill example.

Unsafe condition. The possible reasons for not clearing up the oil spillage were lack of time and not seeing it as part of the job. Asking 'why?' about the lack of time could produce the following types of answer:

- management pressure
- piece work
- wanted to get home
- understaffing
- and so on.

Again, the remedial action suggested will depend on the answer obtained. There is little point in suggesting that people take time to clear spillages if management are continuing to insist on giving production priority.

Unsafe act. Possible reasons for the spillage of oil may include using inappropriate implements and using an inappropriate system of work. Asking 'why?' about the inappropriate system of work might produce the following types of answer:

- no one has prepared a system of work
- the people who do the work do not know about the system of work
- the recommended system of work is impractical
- the recommended system of work is out of date
- and so on.

As before, whichever reason is identified, it should be followed up so that any remedial action suggested is as relevant and practical as possible.

Lack of management controls. Possible reasons for the lack of management controls were the pressure for production, lack of funding and failure to produce written systems of work. Asking 'why?' about written systems of work might produce the following types of answer:

- no one knows it is necessary
- no one has the time
- no one has the skills
- no one has clear responsibility
- and so on.

It should be noted that as the basic investigation moves from the loss domino to the lack of management controls domino, a wider range of people will have to be interviewed. The injured person, for example, is unlikely to have the required information on lack of management controls. He or she can probably tell you about the effects of lack of controls but is unlikely to know the reasons why the controls are not in place.

Identifying who should be interviewed in the course of an investigation and knowing which questions to ask are matters of experience and practice and, as with the other skills aspects of accident investigation, they should be practised whenever possible.

Note also that if a manager is conducting an investigation into an accident within the area of another manager's control, a conflict of interests may arise. The person who should be implementing management controls may have a tendency to avoid going into details of the weaknesses in management control as thoroughly as might be required. In these circumstances it may be preferable to hand the investigation over to, or seek the assistance of, a neutral investigator.

Safety professionals have a related problem when there are requests from managers to take part in investigations. In general this is to be encouraged since it increases management involvement in safety matters, but it should be explained to these managers that they may have to be interviewed as part of the investigation if lack of management control is identified as an underlying cause.

3.7.4 Writing investigation reports

It is not always necessary to prepare a formal written report of an accident investigation, but where it is, the techniques of good report writing should be followed. Key points on report writing are given below.

It is preferable to prepare a draft report since this provides an opportunity to check that nothing has been omitted from the investigations. In particular, that the information is available for:

- making any statutory or other notifications (this aspect was dealt with earlier in this chapter)
- making reasoned suggestions on measures for preventing recurrence, and
- any other tasks, e.g. completing insurance claims.

Some people find that drafting reports is best done with techniques such as system diagrams, system maps and flow charts and it is worth experimenting with these techniques to find out their suitability.

The first question to ask about a final report is whether or not it is necessary. Answers should be based on whether there is an audience, who they are and what this audience needs from the report. In many cases, a detailed draft report is adequate as a record, and as a basis for justifying remedial actions.

The following points should be covered where a final report is required:

(i) *Good signposting.* Any report, but especially a long one, will be difficult to read and action if the various sections are not clearly identified. If the report is intended for more than one audience, the sections relevant to particular audiences should be clearly identified.

(ii) *Separate fact and opinion.* Facts should be unarguable, opinions can, and should, be debatable. It is good practice to keep the two separate.
(iii) *Base opinions on the facts.* Conclusions should not be drawn which cannot clearly be supported by the facts presented, nor should conclusions be drawn which do not take all of the relevant facts into account.

3.7.4.1 Feedback of investigation results

The relevant results of investigations, including any recommendations for remedial action, should be fed back to all of the people who were involved in the investigation.

If this is not done, there may be the following detrimental effects:

- Subsequent investigations will be more difficult, and less information will be given, because people will have seen no results from helping with earlier investigations.
- Credibility will be damaged since people will have been told that the investigation is to prevent recurrence and they have received no instructions on the action to take.

Even though the results from a particular investigation indicate that no action needs to be taken, the results and the reasons for taking no action should be fed back to those who were involved.

3.7.5 Learning from minor incidents and near misses

It is often the case that only the more serious incidents are considered worthy of investigation. The rationale for this is usually that investigations take time and, therefore, cost money so that they are only worth doing when there has been a significant loss. However, various researchers have demonstrated that there is no relationship between the causes of accidents and the seriousness of the outcome and that, for example, minor injuries have the same range of causes as major injuries.

It follows from this that as much can be learned from investigating individual minor incidents and near misses as can be learned from investigating individual major injuries. Since it is also the case that there are many more minor incidents than major incidents, investigation of minor incidents gives us many more opportunities to learn from what has gone wrong.

Since there are so many minor incidents, we are left with the practical problem of the time required for adequate investigation of them all. There are two ways of dealing with this problem.

1 Provide managers with the competences to carry out proper investigations so that the required work is spread among a number of competent persons.

2 Identify patterns in minor incident occurrence and investigate groups of minor incidents which are likely to have related causes. How this pattern identification is carried out was described in section 3.6 on epidemiological analysis.

Assuming that managers can be trained in investigation techniques, the first option is to be preferred. However, the second option can provide an acceptable alternative and it should be used as a backup when managerial investigations are in place.

3.8 The use of computers

3.8.1 Introduction

This section consists of a brief description of the sorts of computer software which are available for the recording and analysis of accident and incident data, and for a range of related data handling tasks. The criteria to be used in selecting software are also briefly discussed.

3.8.2 Hardware and system software

There are many types of computer (usually referred to as hardware) but the most common type is the personal computer (PC), either in its desktop form, or as a portable. This discussion will, therefore, be restricted to software available on PCs.

Before any application program can be run on a PC it has to be equipped with system software. This software does a number of things but essentially it is an interface between the hardware and any application program to be run. The major practical value of system software is that people who write, for example, statistical programs do not have to produce a different version for each different type of hardware. Instead, they write a program for a particular type of system software. The most common system software is Windows in its various versions, including Windows 95 and Windows 97. This discussion will be restricted to software packages which run under Windows, but many of the points made will also apply to other systems software.

3.8.3 The nature of programs

The sorts of programs discussed all operate in essentially the same way. Each one provides a framework, or shell, into which data can be put and, for the present purposes, the programs can be classified according to the types of data they accept. The main categories are as follows:

- *Free format text, diagrams, pictures, tables, etc.* These data types are all accepted by programs such as word processors, desktop publishing packages and presentation packages.

- *Structured alphanumeric data.* These data types consist of mixed letters and numbers in a highly structured format of records and fields. Database programs accept these data types, including the specialised database programs used for specific purposes such as accident and incident recording.
- *Questions and answers.* This is a subcategory of the structured alphanumeric data but because it has special relevance to health and safety it is dealt with separately. Packages for active monitoring, audit, attitude surveys and measuring safety culture accept these types of data.
- *Numeric data.* Spreadsheets are the most common programs for numeric data but these types of data are also used by the specialised statistical packages.

There is always an overlap between programs, for example word processors will do elementary calculations. However, all programs are designed to deal primarily with a single data type. Specific programs are dealt with after some general points.

In theory, there could be one computer program which did everything but, in practice, the more a computer program does, the more difficult it is to learn and use. For this reason, program authors compromise in two main ways:

1 *Reducing functions.* This involves limiting the number of things the program encompasses, for example the sorts of calculations that can be done using a word processor, or the level of word processing that can be done using a spreadsheet.
2 *Reducing flexibility.* This involves limiting the data the program will accept, or the number of things which can be done with these data. For example, any database program can be used for accident and incident recording but database programs are difficult to learn. A program designed solely for accident and incident data, although it is less flexible, should be much easier to learn and use.

However, the link between functionality and flexibility, and speed of learning and ease of use, depends on the skill of the software designer. Some very limited programs are badly designed and are difficult to learn and use, while some very powerful programs are relatively easy to learn and use.

3.8.4 Free format text programs

The main programs in this category include word processors and presentation packages.

So far as the present purposes are concerned, the primary use for these packages is for getting messages over to other people, either as a written report or as a presentation.

The key point to consider when selecting suitable software of this type is whether it will accept data directly from the other packages being used. Having to retype data, particularly numeric data, is tedious and error prone, and it is preferable to have a word processor and presentation package which will read data directly from the output of the other packages in use.

It is prudent to select well-known packages such as Word for Windows (word processor) and Powerpoint (presentation package) since authors of other software are likely to ensure that the output from their programs will be compatible.

3.8.5 Structured alphanumeric data

The main programs in this category include general databases and databases designed for use with specific types of data such as accident records.

General databases such as Access have a very wide range of functions and are very flexible. However, they are difficult to use without some programming experience or the willingness to devote time to learning how to use them.

There are two separate stages in the use of general databases:

1 Setting up the database so that it will do the recording and analysis required. If, for example, a general database is to be used to record and analyse accident and incident data it would be necessary to set up the fields for recording such things as name of person injured, time of injury and number of days lost. This is specialised work requiring a high level of skill.
2 Entering data into the framework created in step 1. This requires a lower level of skill but, unless step 1 has been carried out properly, it will be highly error prone. For example, step 1 should include building in automatic checks on the data being entered with appropriate error messages when incorrect data are entered.

Because of the high levels of skill required to set up general databases for specific uses, it is not usually worthwhile for health and safety professionals to learn the skills required. What normally happens is that the health and safety professional specifies what is required and then hands over the work of setting up the database to the IT professionals who then produce a program which looks like a specific database when it is being used for data input and analysis.

Specific databases are available for a wide range of uses including the recording and analysis of data on accidents, risk assessments and various test results such as audiometry and LEV tests. Several different versions of each database type, which differ in function, flexibility and price, are available on the market.

The key selection strategies for these types of databases involves two main elements:

1 Being clear about what data are to be recorded and what analyses are to be carried out. Software suppliers will try to convince potential purchasers that their program does what is required, but this is not always the case. On the other hand, purchasing new software should be taken as an opportunity to review what is being done by way of recording and using data since there is little point in computerising a poor paper system.
2 Looking to the long term. Many program demonstrations are carried out with just a few records on a highly specified computer and they appear fast and easy to use. Ask to see demonstrations involving the sort of computer with the numbers of records there will be in the system in two to three years' time. Some programs may be so slow as to be unusable.

The health and safety trade press carries advertisements for these types of specific databases and it is easy to get further information simply by phoning the suppliers.

3.8.6 Questions and answers

The main uses for programs of this type are the recording and analysis of active monitoring data, audit data, and data from surveys such as attitude or safety culture surveys. The strategy for the selection of these programs includes the points already made about specific databases, plus the following:

● *Flexibility of the question set.* Some programs are supplied with a set of questions which cannot be altered, while others can be supplied in a form which allows users to put in their own questions 'from scratch', or tailor a set of questions provided with the program. Fixed questions are fine so long as they exactly meet an organisation's requirements, but this is not often the case.
● *Use of more than one question set.* Some programs allow the use of only one set of questions (fixed or tailored) for all analyses while others allow the use of as many different sets of questions as may be required. The latter type of program is to be preferred when, for example, there is a wide range of risks and it is preferable to avoid asking people a lot of questions which do not apply to them.
● *Analysis options.* Some programs have very limited analysis options while others provide a range of alternatives. An important point to note is the extent to which the program allows 'labelling' of the answers to a particular set of questions. For programs designed for auditing, it may be adequate to have one label for each set of questions, usually the location which was audited. However, for attitude and safety culture surveys a range of labels will be required including, for example, department, level in the management hierarchy, length of time with the company and age.

3.8.7 Numeric data

Programs for numeric data are similar to databases in that they are split into general programs, i.e. spreadsheets, and programs which are designed to do specific things with numeric data, i.e. statistical packages.

So far as spreadsheets are concerned, the principles of their use and selection are the same as for general databases, although people in general tend to be more familiar with spreadsheet use.

There is a range of statistical packages available ranging from cheap and easy to use packages which will do most of the basic statistical tests to expensive, 'heavy weight' packages suitable only for the professional statistician. However, none of these packages will compensate for poor statistical technique. Easy, accurate calculation of confidence limits are of no value if incorrect types of confidence limits are being used.

3.8.8 Choosing software

Summarising the steps to take in choosing appropriate software of any type:

- Know the hardware and system software to be used since this will put restrictions on which programs can be used.
- Know exactly what is to be achieved by using the software. However, always take the opportunity to review the extent of the recording and analysis being carried out since the availability of software may make it possible to do more than is currently being done.
- Check what relevant software is on the market. This is probably best done by reading the health and safety trade press, or one of the many computer magazines.
- Get a demonstration of the software under conditions which match those under which it will be used. Many software houses will supply 'demonstration versions' which can be tried out on the computer setup to be used.
- Do a cost benefit analysis on the options available. It is unlikely that any package will exactly meet requirements but remember that having a program written is likely to be several orders of magnitude more expensive than buying one 'off the shelf'. A decision may have to be made as to whether being able to do exactly what is required is worth the extra cost.

References

1. Health and Safety Executive, Guidance Book No. HSG 65, *Successful health and safety management*, HSE Books, Sudbury (1997)
2. Health and Safety Executive, Guidance Book No. HSG 96, *The costs of accidents at work*, HSE Books, Sudbury (1997)
3. Health and Safety Executive, Legal Series Book No. L 73, *A Guide to the Reporting of Injuries, Diseases and Dangerous Occurrences Regulations 1995*, HSE Books, Sudbury (1996)

4. Moroney, M.J., *Facts from Figures*, Penguin Books (1980)
5. Shipp, P.J., *The Presentation and Use of Injury Data*, British Iron and Steel Association. No date. (Out of print, but copies should be available through interlibrary loan services.)
6. Siegel, S.S., *Non-parametric Statistics for the Behavioural Sciences*, McGraw Hill (1956)
7. Whaler, D.J., *Understanding Variation – the Key to Managing Chaos*, SPC Press,
8. Health and Safety Commission, *Health and Safety Commission Annual Report, Statistical Supplement*, HSE Books, Sudbury (published annually)
9. Hale, A.R. and Hale, M., Accidents in perspective, *Occupational Pschology*, **44**, 115–121 (1970)
10. Appendix 5 of reference 1

Chapter 4

The individual

Professor A. R. Hale

4.1 What is behavioural science?

Behavioural science has three main aims: to describe, to explain and to predict human behaviour. Systematic description is the essential foundation of this, as of any other scientific subject. The safety adviser is interested particularly in behaviour at work and more particularly in the behaviour of people in situations which endanger their health or safety. Even describing behaviour in such situations is not easy; many people describe what they expected to see rather than what they actually did see. Explaining behaviour requires theories about why it happens; we need to get to this level of understanding in order to understand behaviour in accidents and to decide how to design hardware and organisations which will be useable by people and complement their skills. What we really want to do is predict in detail how decisions on selection, training, design and management will influence the way people will behave in the future. This is a very severe test of theories about individual behaviour, and psychology is often not far enough advanced as a science to withstand such scrutiny. In this chapter the aim is to describe in broad terms what is known and can be used to help us understand and guide human behaviour.

A human individual is far more complex than any machine and when individuals are placed together in groups and organisations the interactions between them add many times to the complexity which needs to be understood. Individuals are also extremely adaptable. They change their behaviour as they learn and if they know that they are being observed. Also, each individual is to an extent unique because of their unique experience. Because of all this the behavioural scientists' task can be seen to be daunting indeed. It is also difficult to measure all the factors which may affect an individual's behaviour at any one moment.

The explanations and predictions of behavioural science therefore have wider margins of error than those which can be offered by engineers or doctors. Statements made about behaviour will usually be qualified by words such as 'probably' or 'in general'. Individual exceptions to the predictions will always occur.

Because of its limitations behavioural science is dismissed by some as being no more than common sense dressed up in fancy language. All individuals must have some ability to explain and predict the behaviour of themselves and others, or they would not be able to function effectively in the world. However, most individuals' explanations and predictions are often proved wrong. Behavioural science used in a systematic and rigorous way can always improve on unaided 'common sense'.

Behavioural science commonly works by developing models of particular aspects of human behaviour. These models are inevitably simplifications of real life, in order to make it comprehensible. The models are frequently analogies drawn from other branches of knowledge. They represent the brain as a telephone exchange or a computer, the eye as a camera, etc. Different behavioural scientists use different analogies. This, to some extent, explains why there sometimes appear to be parallel and incompatible theories about the same aspect of human behaviour. Analogies are powerful and useful, but they have limitations which must always be acknowledged. They can never be perfect descriptions of the way that an individual functions, and will be useful only within their limits. In the sections which follow some of the models will be described and used to explain particular aspects of behaviour. Readers are urged to use them, but with care.

4.2 The relevance of behavioural science to health and safety

Here are some of the questions which behavioural science can help to answer:

> What sort of hazards will people spot easily, and which will they miss?
> At what times of day and on what sorts of job will people be least likely to notice hazards?
> Can you perdict what sorts of people will have accidents in particular circumstances?
> Why do people ignore safety rules or fail to use protective equipment and what changes can be made in the rules of equipment to make it more likely that people will use them?
> If people understand how things harm them, will it make them take more care?
> What knowledge do people need to cope with emergencies?
> What extra dangers arise when people work in teams?
> How do company payment, incentive and promotion schemes affect people's behaviour in the face of danger?
> How can training help to make people take care?

Can you frighten people into being safe?
When are committees better than individuals at solving health and safety problems?

The list of questions can go on almost indefinitely. Before studying behavioural science it is a valuable exercise to draw up a list of questions relevant to your own work place. See how many of the questions you have answered, or reformulated at the end of your study.

4.3 The human being as a system

A common model used in behavioural science, and in the biological and engineering sciences, is the 'systems' model. Systems are defined as organised entities which are separated by distinct boundaries from the environment in which they operate. They import things across those boundaries, such as energy and information; they transform those inputs inside the system, and export some form of output back across the boundaries. Open systems are entities which have goals or objectives which they pursue by organising and regulating their internal activity and their interchange with their environment. They use the feedback from the environment to check constantly whether they are getting nearer to or further away from their objectives. *Figure 4.1* shows a generalised system model.

Such models can be applied to a single cell in the body, to the individual as a whole, to a group of individuals who are working together, and to an organisation such as a company.

Figure 4.2 considers the human being as a system for taking in and processing information. Accidents and ill-health are conceived as damage which occurs to the system when one or other part of the system fails. The human factor causes of accidents can be classified according to which part of the system failed. The following sections discuss the various stages in the model set out at *Figure 4.2*.

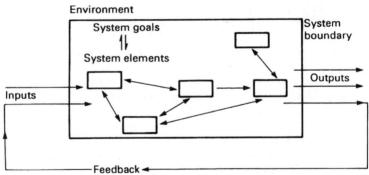

Figure 4.1 Simplified system model. (Adapted from Hale and Glendon[1])

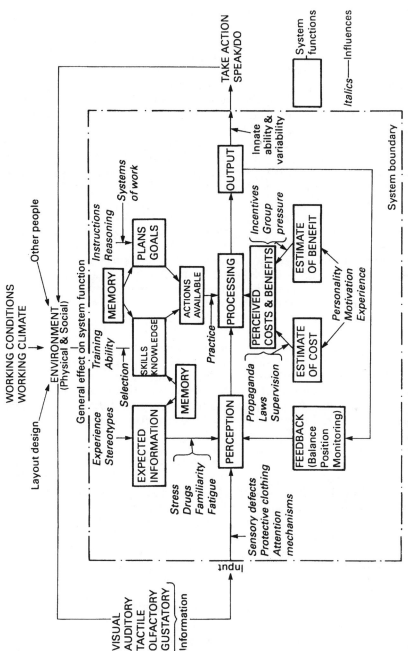

Figure 4.2 Systems model of human behaviour. (Adapted from Hale and Hale[2])

4.4 Facets of human behaviour

4.4.1 Goals, objectives and motivation

Any understanding of human behaviour must start with an attempt to describe the goals and objectives of the human system. Individuals have many goals. Some such as the acquisition of food and drink are innate. Others are acquired, sometimes as means of achieving the innate goals, and sometimes as ends in themselves, for example the acquisition of money, attainment of promotion, purchase of a house, etc. Some are short term, e.g. food at dinner time; others are much longer term, e.g. earning enough for retirement. In some cases the short and long term goals may be in conflict; e.g. a person may fail to check equipment before starting work in order to satisfy the short term goal of getting the job over as fast as possible, as a result jeopardising the long term goal of preserving his own health and safety.

Not all goals are consciously pursued, either because the person may not want to admit even to himself that he is pursuing a particular goal, or because the goal is so basic that it has been built into the person's behaviour and no longer requires any conscious thought.

An individual's goals can be conceived of as vying with each other to see which one will control the system from moment to moment. People will therefore show to some extent different and sometimes contradictory behaviour from day to day, and certainly from year to year, depending which goal is uppermost at the time. However, people will also show consistency in their behaviour, since the power of each of their goals to capture control of the system will change only slowly over the sort of time periods which concern those interested in behaviour at work.

Many theorists have written about motivation, particularly motivation at work. They have emphasised different aspects at different times. The following is a brief historical survey of the main currents of theory.

1 *F. W. Taylor and Economic Man.* Taylor divided people into two groups: potential managers who were competent at and enjoyed planning, organising and monitoring work, and the majority of the workforce who did not like those activities but preferred to have simple tasks set out for them. Taylor considered that, once work had been rationally organised by the former and the latter had been trained to carry it out, money was the main motive force to get more work out of them. His ideas of Scientific Management[3] encouraged the development of division of labour and the flow line process, work study and the concentration on training, selection and study of the optimum conditions for work.

2 *Elton Mayo[4] and Social Man.* Studies in the 1930s at the Hawthorne works of the Western Electric Company in Chicago which set out to discover optimum working conditions led to the realisation that people were not automata operated by money, but that they worked within social norms of a fair day's work for a fair day's pay. It showed that they were responsive to social pressure from their peers, and to interest shown in them by the company. This led to a new emphasis on the role

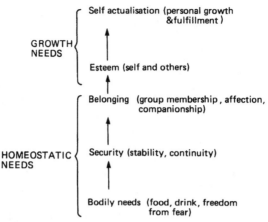

Figure 4.3 Hierarchy of needs. (After Maslow[5])

of the supervisor as group leader, rather than as autocrat, and also to a greater emphasis on building group morale.

3 *Self Actualising Man, Maslow*[5]. Maslow looked at the motivation of people who were successful and satisfied with their work. He found that there was always an important element of achievement, self-esteem and personal growth in their descriptions of their behaviour. He put forward his theory of the hierarchy of needs (*Figure 4.3*) to express this concept of growth. He postulated that the homeostatic needs had to be satisfied before the growth needs would emerge. Although this hierarchy has not been subjected to rigorous scientific confirmation, it is broadly borne out by research studies.

Other theorists who have studied achievement motivation are Atkinson[6] and McClelland[7] who studied the motivation of entrepreneurs compared with government employees, showing a clear difference in the importance they accorded to commercial risk taking and to success.

4 *Complex Man*. Modern motivation theory tries to incorporate what is valuable from all of the earlier theories, and recognises that there are individual differences in the strengths of different motivations both between individuals and over time in the same individual. As far as possible, incentives need to be matched to the individual and the situation (the job of person-oriented man management). It is also recognised that the human system is more complex than many early theories postulated, and that expectations play a strong part in motivation[8]. In other words the force of a motivator is dependent on the sum of the value of the reward and the expectancy that a particular behaviour will lead to the reward. If someone perceives that it will take a great deal of effort to gain any increase in reward, or that the reward does not appear to be dependent upon how much effort is actually put in, their behaviour will not be influenced by that reward.

The unique combination of goals and behaviour which represents each individual's adaptation to the environment in which he finds himself is one definition which is given to the word personality (see section 4.7.1.4). Thus, those who habitually place a high value on their need for acceptance by people around them are called gregarious or friendly, whereas those people who habitually subordinate their need for approval by others to their goal of achieving high status in the organisation, we call ambitious.

It may be thought axiomatic that the preservation of the self (i.e. of safety and health) would be one of the basic goals of all individuals. Clearly this is not a goal of all people at all times, as the statistics of suicides must indicate. Some psychologists[9] have argued that there are deep seated unconscious desires in some people at some times to damage themselves, which constitute a major factor in accident causation. However, research carried out among normal populations who are not receiving psychiatric treatment indicates that there is little, if any, ground to support this view[10]. It is perhaps best to assume that preservation of health and safety is indeed a basic goal of the normal human system, and that most failures to achieve that goal are because individuals do not perceive that their safety is immediately threatened, and so other goals which the individual has are given priority over the one of self-preservation. If risks are perceived to be small and gains great, the individuals are willing to trade off a slight increase in risk for a bigger short term gain in speed or comfort. In exceptional circumstances, such as in saving a child from drowning, personal safety may be put at considerable risk, but this does not often happen otherwise.[11]

4.4.2 System inputs

Information gets into the human system through the sense organs. These have traditionally been considered as five in number – sight, hearing, touch, smell and taste – but touch can be sub-divided into the senses of pain, pressure and temperature, and in addition there is the 'proprio-ceptive' or 'kinaesthetic' sense which transmits information from the muscles and joints to the brain, informing it about their position in space, and their orientation one to another. Closely related to this is the sense of balance.

Hazards which are not perceptible to the senses will not be noticed unless suitable alarms are triggered by them or warnings given of them. Examples are odourless, colourless gases such as methane, X-rays, or ultrasonics, or hazards in the dark. The canary falling off its perch in the mine because of its greater sensitivity to methane was an early example of a warning device, subsequently superseded by the colour change in a safety lamp flame and now the methanometer.

If any of the senses are defective the necessary information may not arrive at the brain at all, or may be so distorted as to be unrecognisable. Some sensory defects are set out in *Table 4.1*. Sensory defects can also be 'imposed' by some of the equipment or clothing provided to protect people against exposure to danger, e.g. safety goggles, gloves or ear defenders.

Table 4.1. Some sensory defects

Sense	Natural and 'imposed' sensory defects
Sight	Colour blindness, astigmatism, long and short-sightedness, monocular vision, cataracts, vision distortion by goggles and face screens
Hearing	Obstructed ear canal, perforated ear drum, middle ear damage, catarrh, ear plugs or muffs altering the sound reaching the ear
Taste and smell	Lack of sensitivity, genetic limitations, catarrh, breathing apparatus screening out smells
Touch senses	Severed nerves, genetic defects, lack of sensitivity through gloves and aprons
Balance	Ménièrés disease, alcohol consumption, rapid motion, etc.

The sense organs themselves have a severely limited capacity for receiving and transmitting information to the brain. The environment around us always contains far more information than they can accept and transmit. There are two types of mechanism which the individual uses to overcome this potential overload of information.

1 Switching or attention mechanisms.
2 Expectancy.

4.4.3 Switching or attention mechanisms

The brain classifies information by its source and type. It is capable of selecting on a number of parameters those stimuli that it will allow through a filter into the system. The setting of the filters on each sensory mode is partly conscious and partly unconscious.

The main visual attention mechanism is the direction of gaze which ensures that the stimulus from the object being looked at is directed to the most sensitive part of the retina (fovea) where it can be analysed in detail. The rest of the field of view is relegated to the less sensitive parts of the retina. In normal activity this centre of focus is shifted constantly in a search pattern which ranges over the field of view until an object of interest is picked out. The visual sense can be tuned to seek out a particular facet such as a defect in a machine or component, provided that we know in advance what characteristics to tune it to. This ability is known as 'perceptual set'.

In the hearing sense there is a similar ability that enables us to tune into the various characteristics of sounds such as direction, pitch, intensity, or timbre. People can therefore 'tune' themselves to pay attention to strange sounds coming from a particular part of a machine, while ignoring all others. It may take several occasions of looking at something before the filter is confident enough that this is what is being sought and allows it through. So we can sometimes look but not see. As well as being variable from time to time perceptual set shows longer term settings which produce differences between individuals because of their interests and their experience; safety advisers notice hazards because they are interested in them and used to finding them; motor cycle addicts spot a Bonneville in a crowded street where others would not even notice that there was a motor bike.

Inputs which do not vary at all are usually not particularly useful to the system, e.g. a constant noise or smell, a clock ticking, the sensation of clothes rubbing on the skin. The filter alters over time to exclude such constant stimuli from consciousness. As soon as they change, however, e.g. the clock stops, the filter lets through this information and we notice it.

These selective attention mechanisms are extremely efficient and invaluable in many tasks. But any mechanism which is selective carries with it the penalty that information which does not conform to the characteristics selected by the filter will not get through to the brain, however important that information is. People can be concentrating so hard on one task that they are unaware of other information. Hence someone can fall down a hole because they were staring at some activity going on in the opposite direction. Presetting the filter can also lead to false alarms; searching a list for the name Jones we can sometimes be fooled by James. The cost of a rapid response is an increase in errors.

4.4.4 Expectancy

Perception is strongly influenced by a source which is located entirely inside the brain. This is expectancy, or the model of the real world which has been built up from experience over an individual's lifetime. This can form a very 'real' alternative to direct input from the world itself. Every one living in an industrial society knows what a motor car looks like and can conjure up a mental picture of one comparatively easily. This means that when confronted with a particular car in the real world there is no need to take in all of the details which are already on file in the brain. The person can concentrate upon only those characteristics which differentiate this car from the 'standard' car of his mental picture, e.g. its colour, or make, or its driver. Again there is a cost. In the UK we are so used to expecting the driver to be sitting on the right of the car that we may not see that this particular car is from abroad and that the person on the left is the one driving.

Machines, processes, people, and whole situations are stored in the brain and can be recalled at will, like files from the hard disc of a computer. This cuts down enormously on the amount of information about any scene which an individual needs to take in in order to perceive and understand it. But where the real world differs from the expectation, problems occur. This is most often the case with situations which go against population stereotypes, for example a machine on which moving a lever downwards turns it off. Other population stereotypes are red for danger and stop, clockwise turns the volume up or shuts the valve. These examples are very widely shared, but in other cases stereotypes for one population may contradict those held by others, e.g. you turn the light on by putting the switch down in Britain but by pushing it up in the USA and parts of Europe. Designs which do not match expectation can trap people into making errors.

Many of the classic illusions seem to be caused by misplaced expectations. For example, if you are sitting in a train at a station and another train alongside moves away, you can experience a sensation that you are moving because that is what you expect. Similarly, sitting in the cinema you believe the voices come from the characters on the screen, not from the loudspeakers at the side.

We tend to perceive pattern, regularity and constancy in the world when it is not really there because we fill out any imperfections with our thoughts, so failing to notice flaws and other irregularities.

In some circumstances these false expectations may result in little more than annoyance and delay. In other cases it may be a prelude to physical damage or injury. For example, a machine operator may reach rapidly towards his pile of components without looking and gash his hand on the sharp edge of one of them which has fallen off the pile and is nearer than he expected; the truck driver may drive rapidly through the doors which are reserved for trucks without looking or sounding a warning because no one is supposed to be there, only to find that someone is using the truck doors as pedestrian access.

The reliance upon expectation is an essential mechanism in skilled operation and many tasks would take a great deal longer to carry out if this was not so. Thought therefore needs to be given to ways in which reality can be made to fit people's models, rather than vice versa. Standardisation of machine controls, layout of work places, colour coding and symbols, etc. are all designed to achieve this result, as are codes of rules such as the Highway Code or Plant Operating Procedures. But standardisation has a hidden snag. The more standard things normally are, the more likely are exceptions to trap someone into an error. So, standards and rules must be enforced 100% to avoid this danger.

Any circumstances which are unclear or ambiguous (e.g. fog, poor lighting) or where an individual is under pressure of time, is distracted or worried, or fatigued, will encourage expectancy errors. In extreme cases individuals may even perceive and believe in what are in fact hallucinations.

4.4.5 Storage

The storage facility of the human system is the memory. The memory is divided into two different types of storage, a long-term, large capacity store which requires some time for access, and a short-term working storage which is of very small capacity and rapidly decays, but can be tapped extremely rapidly.

The short-term memory is extremely susceptible to interference from other activities. It is used as a working store to remember where one has got to in a sequence of events, for example in isolating a piece of equipment for maintenance purposes. It also stores small bits of information between one stage of a process and another, such as the telephone number of a company between looking it up in the directory and dialling the number.

Long-term memory contains an abundant store of information which is organised in some form of classification. Any new information is perceived in terms of these categories (closely related to expectations) and may be forced into the classification system even when it may not fit exactly. In the process it can become distorted. This process probably also results in specific memories blurring into each other, with the result that the wrong memory may be retrieved from the store when it is demanded.

People are not able to retrieve at any one occasion all the things which they have stored in their memory. There are always things which they know, but cannot recall, and which 'pop out' of store at some later stage. They are there but we have forgotten where we put them. This sort of limitation can frequently be overcome by recalling the circumstances in which the original memory was stored, or by approaching it via memories which we know were associated with it. Unavailability of memories may be crucial in emergency situations where speed of action is essential. A technique for overcoming unavailability is to recall and reuse the memories (knowledge and skills) at regular intervals. Refresher courses, emergency drills, and practice sessions all perform this function. However, one unwanted side effect of constant recall and reuse of memories is that they may undergo significant but slow change. When the memory is unpleasant, or shows the individual in a bad light, it is extremely likely that distortion will occur at each recall and it will be these that will be remembered rather than the original story. Testimony following an accident is notoriously subject to such distortion. People can quite genuinely remember doing what they should have done (the rule) rather than the slip they actually made.

4.4.6 Processing

To use information it must be processed. This may be done 'on-line' or 'off-line'.

4.4.6.1 On-line processing. Routines and skills

This is the moment-to-moment decision making about what action to take next in order to cope with and respond to the environment around the individual. This processing has to be funnelled through a narrow capacity channel which can only handle small numbers of items at one time. This limited capacity can be used to best effect by grouping actions together as packages or habits which can be set in motion as one, rather than as separate actions. Such habits form the basic structure of many repetitive skills, for example signing one's name, loading a component into a machine, changing gear in a car. Such grouping of activities does, however, carry with it the penalty that, once initiated, the sequence of actions is difficult to stop until it has run its course. Monitoring is turned down low during the routine. This can result in injury, for example if someone steps off a loading platform at the point where the steps always used to be, without remembering that recently they have been moved. The packaging of actions in these chunks also places greater premium on correct learning in the first place, since it becomes very difficult to insert any new actions into them at a later stage (see section 4.7.4.1).

4.4.6.2 Off-line processing. Decision making and intelligence

This is the facility whereby people can simulate in their mind the results of different possible courses of action before they make any decision about which course to choose. This skill is an immensely valuable one because it allows some courses of action to be rejected without ever trying them on account of the unpleasant consequences which we correctly predict. However, as a skill it depends upon knowledge of how factors interact and the ability to manipulate many factors together in the mind. This in turn is related to intelligence, and to the amount of practice in using the skill.

The ability to learn, to manipulate concepts in the head and to solve problems is one way of defining intelligence or cognitive efficiency. Psychological research is full of conflicting views of exactly how intelligence should be defined, whether it is largely innate or can be modified significantly by environment, and how it should be measured. This conflict arises from the complexity of obtaining evidence with which to support or disprove the various theories. Intelligence testing was first carried out by child psychologists – notably Binet in the first years of this century and developed through its massive use by the American army in the first world war. Binet[12] invented the term IQ (intelligence quotient) to describe his finding that, in children, greater intelligence seemed to result in a child being able to do certain tasks sooner than the average child.

$$\text{Intelligence quotient, IQ} = \frac{\text{mental age} \times 100}{\text{chronological age}}$$

Therefore by definition an average IQ is 100. Most people who gain a university degree or equivalent professional qualification have an IQ of more than 120, while an IQ of 70 or less will make all but simple routine work beyond the person's intellectual capacity.

Most theories of intelligence agree that it can be subdivided into special aptitudes e.g. verbal, numerical, spatial, manual, mechanical, musical. These represent the sort of problems that a given individual is best at solving in the head. Tests can be found for all these aptitudes. However, they need to be selected, administered and interpreted by experts if they are to be valid and useful.

Errors in off-line processing usually relate to attempts to simplify the decision so that it can be handled in the head. Such simplifications involve ignoring the effect of some factors, rejecting out of hand some courses of action without considering them, and limiting the degree to which the consequences of a course of action are thought through before making a decision. All these limitations, which may be unconsciously imposed, can result in unsafe decisions being made.

Individuals are also not entirely logical in the way in which they make decisions. The value assigned to particular outcomes, such as amount of effort saved, money earned, approval obtained or withheld by colleagues and superiors, etc. is a subjective one. It will be influenced by the personality of the individual making the decision, and by experience of the way in which previous decisions have turned out. Small probabilities are consistently poorly assessed and people rely upon such illogical factors as 'luck', which they believe they can influence, rather than accepting that some things such as roulette wheels or dice are entirely random machines.

The rate and efficiency of mental processing are also limited by the level of arousal of the brain. At low arousal levels performance is poor, rising with increased arousal to an optimum and then falling with further increases. This change in arousal corresponds roughly to a movement from drowsiness, through optimum coordinated performance, to the confused activity resulting from over-anxiety and panic.

4.4.7 Output

Once the decision to act has been made, the remaining limitations on the human system are those of its capacity to act, e.g. its speed, strength, versatility. Humans differ from machine systems in that their actions are not carbon copies of each other, even when the individual is carrying out the same task again and again. The objective may be unchanging, but the system adapts itself to small changes in body position, etc. to carry out a different sequence of muscular actions each time to achieve that same result. This use of constantly changing combinations of muscles in coordination is an essential means for the body to avoid fatiguing any particular muscle combination.

In all human actions there is also a trade-off between the speed of an action and its accuracy. Speed can be improved by reducing the amount of monitoring which the brain carries out during the course of the action, but only at the cost of reducing the accuracy.

4.4.8 Effects of the environment and degradation of performance

All of the above limitations have been described in isolation from the effects of the external environment. General environmental conditions such as noise, glare and lighting level, dust and fumes, social environment, etc. will influence the factors which have been described above. For example, noise and high temperature both have an effect on the arousal level. Noise increases it, heat decreases it, and both have an effect on the accuracy of detection of information and the speed of processing it. These physical environmental factors are dealt with in the chapters on Occupational Health and Hygiene. The effects of fatigue and the social environment upon individual behaviour are dealt with here.

The performance of the human system is only at an optimum within certain environmental limits. As part of the price of its sensitivity and flexibility the human system is susceptible to the influence of a very large range of factors which can affect its performance. Unlike machines, human beings show a slow and often subtle degradation of performance over a wide range of environmental conditions, but arrive at a total breakdown only comparatively rarely. This means that individuals can maintain some sort of functioning long after they have passed the peak of their performance, but it also blurs the point at which they should stop in order to avoid errors. Regulations and good practice on working hours for coach drivers, hospital doctors and others must wrestle with the problem of making these black and white decisions at some point of the continuum of shades of grey.

Performance is degraded under the following types of situations:

1 Working for too lengthy a period, which produces fatigue.
2 Working at times of day when body mechanisms are not functioning efficiently, i.e. the diurnal rhythm is disturbed.
3 Loss of motivation to perform.
4 Lack of stimulation resulting in lowered arousal.
5 Working under conditions of conflict, threat, both physical and psychological, or conditions which threaten the body's homeostatic or coping mechanism and cause stress.

4.4.8.1 Length of working periods and fatigue

The 1833 Factory Commission[13] said in its report that the then existing hours of work of children led to 'permanent deterioration of the physical constitution and the production of various diseases often wholly irremediable, and exclusion from the means of obtaining education, elementary and moral, and of profiting by those means by reason of excessive fatigue'. This early quotation underlines the fact that the length and distribution of working hours have social and physical effects, both of which must be borne in mind when considering permitted or agreed working hours. Long periods of working produce:

1 Muscular fatigue resulting from overloading of individual muscle groups, either through static loading to maintain a posture, or through awkward or repetitive dynamic loading. In addition to limiting working hours, the cure to this problem lies in design of work places to minimise static working load and to allow for the utilisation of the most efficient muscle groups, and the opportunity to rest muscle groups by shifting posture.
2 General mental fatigue characterised by an increase in the length and variability of reaction time, especially for decision making. This leads to an increase in errors and a tendency to neglect peripheral aspects of tasks such as checking routines. These effects can be demonstrated in most tasks after periods of uninterrupted working of between 10 and 50 minutes, depending upon the task load. Rest pauses of one or two minutes interposed when the performance begins to fall from its

optimum level are sufficient to restore functioning to its former level. If performance is allowed to carry on without a break until more obvious signs of degradation have appeared, then proportionately longer rest pauses are required for complete recovery. If the task is machine paced, or strong motivation from pressure of work or incentive bonus schemes prevents natural breaks, artificial breaks should be introduced in order to maintain performance at an optimum level.

If the working day exceeds about ten hours including overtime there is a highly significant reduction in output, and increased risk of accidents on many tasks. Depending upon the intensity of work, the type of work load and any loads imposed by activity outside working hours this effect may occur earlier. It occurs with mental work as much as with manual and physical work[10].

4.4.8.2 Distribution of working hours[14]

Around 10% of the working population spend some time working on a night shift, and a larger proportion who are on shift work, work outside the period 7.00 a.m. to 1.00 p.m.

Body systems follow a cyclical variation in activity which is linked to the 24 hour light-dark cycle. This rhythm is known as the diurnal or circadian rhythm. *Figure 4.4* shows the diurnal rhythm for performance on simple mental tasks. The rhythm for other activities such as motor skills, or tasks involving short-term memory may be out of phase with this, resulting in performance on different tasks being best at different times of day. The difference between performance at the peak and the trough of the curve is of the order of 10%, which is as significant as the degradation in performance caused by a blood alcohol level at the legal limit or by approximately two hours' loss of sleep on the previous night.

The rhythms are keyed into the sleeping/waking cycles as well as the light/dark cycle. These normally work together, but if people work at night, rhythms are thrown into some disarray and take time to begin to

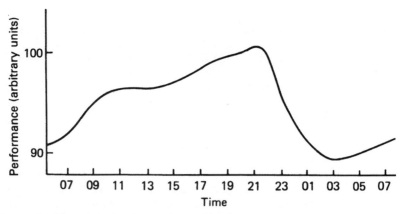

Figure 4.4 Diurnal rhythm for simple mental tasks

adjust. Adjustment begins to be apparent after 2–3 days, and goes on increasing up to a period of about 14 days provided that the person continues both to live and work on a night-time schedule, and does not return to day-time living over a weekend. Even after two weeks the curves have not fully reversed, but have flattened out.

Night workers have the additional disadvantage that they are trying to sleep when the rest of the world is awake and making a noise. Hence their sleep is far more disturbed than that of day workers. Night and evening workers also suffer from a major disruption to social life. This can result in conflicts and stress generated within the family. Studies of night workers show that they tend to have a higher incidence of gastro-intestinal disease such as ulcers, and nervous disorders.

There is no clear evidence that the physical and health effects on women and young persons are greater than on men. The original reasons for the introduction of the ban on night work for protected persons were as much for its supposed moral dangers as for its health effects.

Many firms cope with the problems of shift and night work by adopting rotating shift patterns. Research on the physical effects of rotation show that the traditional British shift pattern of a weekly rotation period is the worst possible compromise, since it results in a constant semi-adaptation and de-adaptation of the diurnal rhythms. Regimes of permanent night work or very rapidly rotating shifts with no more than two nights on night work are far better from these points of view. However, social factors often make the weekly rotation pattern more acceptable to those who have to work shifts.

4.4.8.3 Stress

To the psychologist stress is the result of an internal or external disturbance which threatens the balance of the system. Causes of stress, named stressors, can be divided into two main groups:

1 Environmental stressors such as extremes of temperature, noise, vibration, bodily injury, hunger, etc. all have both physiological effects as well as psychological effects.
2 Psychological stressors such as threat, the inability to achieve valued goals, conflict, physical or social isolation, intense periods of mental activity or excitement, or dramatic changes in life style, e.g. retirement, redundancy, marriage.

There is extremely wide variation in the reaction of different individuals to the same stressor. In the case of physiological stressors this variation is largely a case of differences in physical or physiological tolerance. In the case of psychological stressors there is the important intervening variable of the perception of the stressor and the degree to which it is seen as a threat to valued goals. Individuals have different abilities to deploy coping responses when faced with a stressor which will not go away. *Figure 4.5* summarises these factors.

The physical response of the body to stressors is to increase the activity of the sympathetic nervous system, and to increase the secretion of the hormone adrenalin. This mobilises the body in the primitive

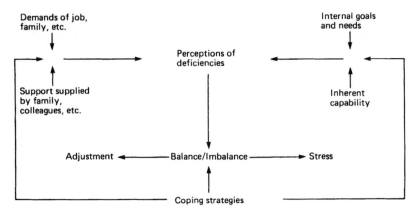

Figure 4.5 Stress components. (Modified from Mackay and Cox[15,16])

condition for fight or flight by increasing muscle tension and heart rate, diverting the blood from inessential activities such as digestion to the muscles and the brain, mobilising stored food reserves, etc. However in modern industrial society physical activity is often not an appropriate response to stressors such as impending examinations, or an angry boss. These physical preparations if maintained for long periods of time can result in harmful side effects such as gastric ulcers, high blood pressure and arteriosclerosis.

The psychological symptoms of stress are disturbed concentration, impaired memory, impaired decision making, tension and aggression, sleep disturbance, and, in severe cases, mood change.

There are a number of coping strategies which can be adopted by individuals. They can withdraw from the source of the stress either physically by leaving their job or going absent, or psychologically by lowering their ambitions, e.g. ceasing to fight for promotion when promotion prospects are blocked. Companies can remove some of the unnecessary demands of the work place, e.g. by work restructuring or by providing greater support through discussion groups, meetings, or counselling services. Finally, there has been some success in bolstering the individual's own resources for countering stress through the teaching of relaxation techniques, through counselling and psychotherapy. A coping strategy open to the employing organisation is to identify people who would appear to be susceptible to stress and to redeploy them into jobs where the demands are low. Some activities such as drinking, smoking or use of tranquillisers which start as coping strategies can end up as health problems in their own right if they result in addiction.

4.5 Types of error

The description of behaviour given up to now has blurred a distinction which is vital in understanding the types of error which people make.

This distinction has arisen from the work of Reason[17] and Rasmussen[18]. They distinguish three levels of behaviour which show an increasing level of conscious control:

1 Skill-based behaviour in which people carry out routines on 'automatic pilot' with built-in checking loops.
2 Rule-based behaviour in which people select those routines, at a more or less conscious level, out of a very large inventory of possible routines built up over many years of experience.
3 Knowledge-based behaviour where people have to cope with situations which are new to them and for which they have no routines. This is a fully conscious process of interaction with the situation to solve a problem.

As a working principle we try to delegate control of behaviour to the most routine level at any given time. Only when we pick up signals that the more routine level is not coping with do we switch over to the next level (see *Figure 4.6*). This provides an efficient use of the limited resources of attention which we have at our disposal, and allows us to a limited extent to do two things at once. The crucial feature in achieving error-free operation is to ensure that the right level of operation is used at the right time. It can be just as disastrous to operate at too high a level of conscious control as at too routine a level.

Each level of functioning has its own characteristic error types, which are described briefly in the following sections.

4.5.1 Skills and routines

All routines consist of a number of steps which have been highly practised and slotted together into a smooth chain, where completion of one step automatically triggers the next. Routine dangers which are constantly or frequently present in any situation are (or should be) kept under control by building the necessary checks and controls into the routines as they are learned. The checks still require a certain amount of attention and the comparatively small number of errors which occur typically at this level of functioning are ones where that attention is disturbed in some way.

1 If two routines have identical steps for part of their sequence, it is possible to slip from one to the other without noticing. This nearly always occurs from the less frequently to the more frequently used routine; for example arriving at your normal workplace rather than turning off at a particular point to go to an early meeting in another building. Almost always these slips occur when the person is busy thinking about other things (e.g. making plans, worrying about something, under stress).
2 If someone is interrupted half way through a routine they may return to the routine at the wrong point and miss out a step (e.g. a routine

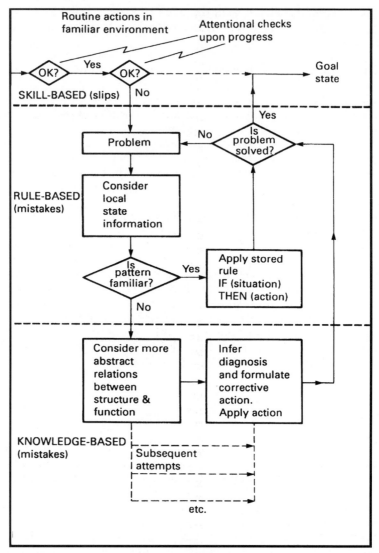

Figure 4.6 Dynamics of generic error-modelling system (GEMS). (From Reason[17])

check) or carry out an action twice (e.g. switching off the instrument they have just switched on because both actions involve pushing in the same button).

3 The final problem at this level is that routines are dynamic chains of behaviour and not static ones. There is a constant tendency to streamline them and to drop steps which appear unnecessary. The most vulnerable steps are the routine checks for very infrequent problems in very reliable systems. (e.g. checking the oil level in a new car engine).

Many of these errors occur because the boundary between skill-based and rule-based activity has not been correctly respected.

These sorts of error will be immediately obvious in many cases because the next step in the routine will not be possible; the danger comes when the routine can proceed apparently with no problem and things only go wrong much later. The cure for the errors does not lie in trying to make people carry out their routines with more conscious attention. This will take too long and so be too inefficient, and will be subject over a short time to the erosion of the monitoring steps. It lies to a great extent with the designer of the routines (and so of the apparatus or system) to ensure that routines with different purposes are very different so that unintended slipping from one to the other is avoided. Where this is not possible extra feedback signals can be built in to warn that the wrong path has been entered by mistake (see 4.6.1.1).

The second line of defence is to train people thoroughly so that the correct steps are built into the system, and then to organise supervision and monitoring (by the people themselves, their work or reference group, and supervisors or safety staff) so that the steps do not get eroded.

4.5.2 Rules and diagnosis

When the routine checks indicate that all is not well, or when a choice is needed between two or more possible routines, people must switch to the rule level. Choice of a routine implies categorisation of the situation as 'A' or 'B' and choice of routine X which belongs to A or Y which belongs to B. This is a process of pattern recognition. This is analogous to computer programme rules of the form IF . . ., THEN . . .

The errors which people make at this level are linked to a built-in bias in decision making. We all have the tendency to formulate hypotheses about the situation which faces us on the basis of what has happened most often before. We then seek evidence to confirm that diagnosis rather than doing what the scientific method bids us and seeking to disprove the hypothesis. This means that people tend to think they are facing well-known problems until they get unequivocal evidence to the contrary. The Three Mile Island disaster was a classic case where operators persisted with a false diagnosis for several hours in the face of contradictory evidence until a person coming on shift (and so without the perceptual set coming from having made the initial diagnosis) detected the incompatibility between the symptoms and diagnosis.

The solution lies in aiding people to make diagnoses more critically (e.g. checking critical decisions with a colleague or supervisor before implementing them).

4.5.3 Knowledge and problem solving

When people are facing situations they have no personal rules for, they must switch to the fully interactive problem solving stage, when they have to rely upon their background knowledge of the system and principles on which it works to derive a new rule to cope with the situation. There are meta-rules for problem solving which can be taught (see section 4.7.4.3). Besides these there is the creativity and intelligence of the individual and the thoroughness of their training in the principles underlying the machine or system. Errors at this level can be traced to:

1 Inadequate understanding of these principles (inadequate mental models).
2 Inadequate time to explore the problem thoroughly enough.
3 The tendency to shift back to rule-based operation too soon and to be satisfied with a solution without checking out the full ramifications it has for the system.

The first two are typical errors of novices, the last more of the expert. Experts are by definition the most capable of functioning at this level, but also the people who need to do so least often, because they have learned to reduce most problems to rules. They may also become less willing to accept that there are situations which do not fit their rules. Almost all experts overestimate their own expertise.

4.6 Individual behaviour in the face of danger

Hale and Glendon[1] combined the insights of the two major models presented so far (*Figures 4.2 and 4.6*) with other sources[19,20] into a model of individual behaviour in the face of danger (*Figure 4.7*). Their model allows us to discuss a number of practical issues in which we are trying to influence human behaviour, e.g. through task design and training.

Danger is always present in the work situation (as in all other situations). The task of the individual is to keep it under control; to avoid errors which provoke an increase in danger and to detect and avoid or recover from danger increasing from other reasons. Much of this activity occurs at the skill- and rule-based levels by more or less routine reaction to warning signals. Only occasionally do people in their normal work situations need actively to contemplate danger. However these occasions are vitally important when they do occur. Examples of such activities are:

1 Designers making decisions about machine or workplace design, plant layout, work procedures, etc. They need to predict the actions of the people who will use these products and the hazards which will arise in use.
2 Operators, safety committees, safety advisers and factory inspectors carrying out hazard inspections, safety audits and surveys, who need to seek out hazards.
3 Policy makers in industry and government deciding whether a level of risk associated with a technology or plant location is to be accepted. Members of the public assessing whether that policy decision is acceptable to them.
4 Planners designing emergency plans for reacting to disasters.

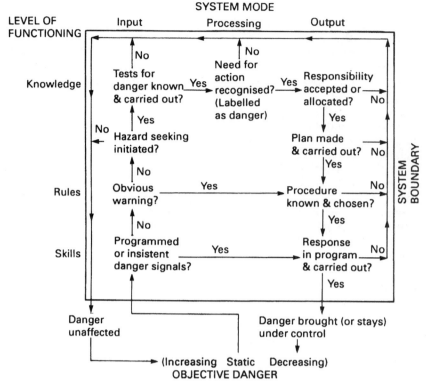

Figure 4.7 Behaviour in the face of danger model[1]

Such decisions and activities are all largely carried out at the knowledge-based level and the borders with the rule-based level.

4.6.1 Hazard detection

This is important in three situations:

1 Detection by operators in routine tasks of deviations from normal which could lead to harm. This is largely skill-based.
2 During formal or informal inspection in the ordinary work environment. This is largely rule-based.
3 Prediction of danger at the design and planning stage of a system. This is largely knowledge-based.

Surprisingly little scientific study has been made of how hazard recognition operates in any of these situations and what alerts people to the presence of danger. What follows is a summary of the available information[1].

4.6.1.1 Routine hazard detection and warnings

Implicit in the structure of skills is a built-in monitoring loop which detects deviations from the programme at the skill level. Deviations from the plan at the rule level are less automatically picked up and depend on good warnings, upon building in physical or procedural blocks which do not allow you to proceed further after such a mistake, or upon intervention by others.

Designers should improve upon the availability of information about deviations and consciously build into their designs feedback about the actions which the individual has just taken, and their consequences. Examples are displays on telephones which show the number you have just keyed in, a click or bleep when a key is pressed hard enough to enter an instruction on a keyboard, commands echoed on a visual display as they are entered on a keyboard, and the use of tick boxes on a checklist to indicate the stage in the check reached.

The following common sense criteria can be used for the design and placing of warnings. They should:

- be present only when and where needed,
- be clearly understandable and stand out from the background in order to attract attention,
- be durable,
- contain clear and realistic instructions about action, and
- preferably indicate what would happen if the warning is not heeded.

Warnings should preferably not be present when the hazard is absent, otherwise people will soon learn that it is not necessarily dangerous in that area. They will then look for further confirmatory evidence that something really is a problem before taking preventive action. The philosophy of 'if in doubt put up a warning sign' is counterproductive unless an organisation is prepared to go to great lengths in enforcing it even in the face of the patent lack of need for the precautions at some times.

If an alarm goes off and there proves to have been no danger, there will be a small, but perhaps significant loss of confidence in it. If false alarms exceed true ones, the first hypothesis an individual will have when a new alarm goes off is that it is a false one. Tong[21] reports that less than twenty per cent of people believe that a fire alarm bell going off is a sign that there really is a fire. The rest interpret the bell in the absence of other evidence as either a test, a faulty alarm or a joke. The recognition of the presence of fire is therefore often delayed, and the first reaction to warnings of a fire is often to approach the area where it is, in order to find out more, rather than to go the other way.

Because warnings must be understood quickly and sometimes under conditions of stress, there can never be too much attention to their ease of comprehension. The language used must be consistent, whether verbal or visual (e.g. a black triangle round a yellow sign always meaning a warning, a red circle with a bar a prohibition). The language must be taught. Written warnings must take account of the reading age of the intended audience and the proportion of illiterates or foreign nationals with a poor understanding of the language. A word such as 'inflammable' should, for example, be avoided because many misunderstand it to mean 'not flammable' (by analogy with 'inappropriate' or 'incomprehensible').

4.6.1.2 Inspection and hazard seeking

During workplace safety inspections the people concerned are already alerted to the possibility that there are hazards present, and are actively seeking them. But to seek is not always to find. Untrained inspectors characteristically miss hazards which have one or more of the following features:

1 Not detectable by the unaided eye, but requiring active looking in, behind or under things, rattling guards or asking questions about the bag of white powder in the corner.
2 Transient, e.g. most unsafe behaviour which can only be discovered by asking questions and using the imagination.
3 Latent; i.e. contingent upon other events, such as a breakdown, a fire, or work having to be done by artificial light.

Inspection must be an active and creative search process, of developing hypotheses about how the system might go wrong. It requires the allocation of time and mental resources. Checklists can help to make the search systematic and to avoid forgetting things, but they should not be allowed to become a substitute for active thinking.

4.6.1.3 Predicting danger. Techniques for human reliability assessment

Prediction at the design stage is an extension of the problem of inspections, made more difficult because there may be no comparable system in existence from which to learn. Imagination and creativity are therefore relevant attributes for the risk analyst in addition to both plant knowledge and expertise in behavioural sciences.

There are large individual differences in how good people are at imagining the creative misuse that operators will make of their systems. Those who are good are known as 'divergent' thinkers. There is some evidence that people who gravitate towards the sciences, mathematics and engineering are more 'convergent' in their thinking, and tend to be more bound by experience and convention, than those who choose social sciences and the arts. They may therefore be less able to anticipate the more unusual combination of events which could lead to harm. This suggests the need for teamwork in design and hazard prediction.

Risk assessment techniques such as HAZOP, design reviews and fault and event trees are systematic methods to guide and record the process of creative thinking. They are also often used to quantify the chance of failure. That step is only legitimate if we are certain that all models of failure have been identified. That is particularly difficult with human-initiated failures because people can act (and fail) in so many more ways than hardware elements. Human reliability assessment techniques should be seen within the framework of the following list of steps[22,23].

1 Familiarisation;
    Plant visit
    System analysis review
    Information gathering
2 Qualitative analysis:
    Task analysis and talk-through
    Performance objectives definition
    Performance situation specification
    Modelling of human performance
    Identification of potential human errors
3 Quantitative analysis:
    Determination of basic error probabilities
    Identification of performance shaping factors and dependencies
    Allowance for recovery factors
    Calculation of final human error contribution
4 Incorporation in Probabilistic Risk Assessment:
    Sensitivity analysis
    Combination with hard- and software failure probabilities

A systematic task analysis is essential for a good prediction of human error. The data for such an analysis come partly from logical analysis of what should happen and partly from observation of what does happen. Such techniques are dealt with in Chapter 2.

Task analysis forms the basis for the use of techniques for error prediction, which are currently in an early stage of development. They all depend upon one or other sort of checklist. For example each sub-task can be subjected to the following standard list of questions to specify what would happen if these types of error occurred and how such an error could happen (cf. HAZOP):

- sub-task omitted
- sub-task incorrectly timed:
  - too soon
  - too late
  - wrong order
- inadequate performance of sub-task:
  - input signals misinterpreted (misdiagnosis)
  - skills not adequate
  - tools/equipment not correctly chosen
  - procedure not correct
  - inappropriate stop point
    - too soon
    - too late
    - not accurate enough
    - quality too low
- routine confusable with other routines

Other checklists[24] have been produced linked to the models of Reason[17] and Rasmussen[18] or derived from ergonomics[25].

Kirwan[26] carried out an evaluation of five techniques of this sort to see which gave the best results when used by safety assessors on four representative tasks. No one technique came out as best in all circumstances. It is a question of horses for courses. Whatever the technique used it must distinguish clearly between:

- The external error mode, which is the effect the human error has on the system (e.g. fails to turn switch, misreads dial, etc.). This is important in linking human error analysis to other parts of the risk assessment.
- The error cause, which is the psychological description of why the error is made. This relates to prevention measures which can be taken.
- The system factors which affect performance on a range of tasks (the Performance Shaping Factors).
- The possibility of error recovery because the initial error is recognised.

Human error quantification is one of the most bitterly disputed areas of risk analysis. Risk analysts starting from the reasonably successful experience of quantifying hardware reliability try to treat human reliability in the same way, so that it can be integrated into their fault tree and event tree analyses. Psychologists doubt the possibility of doing this because:

- Human behaviour is far more complex than component behaviour and can fail in many different ways in different circumstances. The data collection problem is therefore enormous.
- Human failure rate is dependent upon the way in which each individual interprets the task and adapts it or themselves to prevailing circumstances.
- Inter-individual differences in error rate are large (typically a factor 3). Intra-individual differences can be considerable over time because of learning processes, working situation, fatigue etc.
- The initial error rate of humans is high, but they detect a great majority of their own errors and failures and correct them.

Kirwan[26] reviews eight techniques for human reliability quantification. Again his conclusion is that no one method stands out above all others and that there is room for much work to develop better methods.

4.6.2 Knowledge of causal networks

Hazard detection has been shown above to be dependent on the mental models people have of the way in which events happen and systems develop. If these mental models are incomplete or wrong they can lead to inappropriate behaviour in the face of hazards. Interview studies[1] show that such problems frequently occur, particularly in relation to occupational disease hazards. Examples of such significant inaccuracies are:

- Men sawing asbestos cement sheets who said that they only wore their face masks when they could see asbestos particles in the air. (Yet it is the microscopic, invisible particles which are the most dangerous because they are in the size range which can penetrate to the lung.)
- Wearers of ear defenders who fail to incorporate the notion of time weighted average exposure in their concept of what constitutes dangerous noise. Hence they fail to realise that 'just taking the ear-muffs off for a few moments to let the ears breathe' in high noise areas can negate much of their protective effect.
- Misconceptions about the link between posture and musculo-skeletal damage such as considering postures as 'relaxed' and therefore good when they showed the body and notably the spine in a slumped position (which in fact puts extra load on the back muscles to stabilise the spine in that position).

Such misconceptions must be put right by training.

4.6.3 Reactions to perceived risk

The reactions of different groups to risks they perceive have been the subject of much research in the past two decades[1,27,28]. Much of it has concentrated on decisions about siting hazardous plants or developing technologies such as nuclear power. A main focus has been the question of 'acceptability of risk'. This term has led to much confusion because it implies that people are, or should be content, or even actively happy with a particular risk level. The word 'accepted' or 'tolerated'[29] gives a better assessment of the situation, since it carries with it an idea that the opportunity to do something about the hazard is a relevant factor in any decision. There is also overwhelming evidence that people do not consider the risk attached to an activity or technology in isolation from the benefits to be gained from it[30]. Therefore no absolute 'acceptable' level for a wide range of different hazards can be meaningful, since the benefits which go with them will vary widely.

A clear distinction has emerged from the research between threats to personal safety, threats to health and threats to societal safety[31]. The factors which people use in assessing each of them appear to weigh differently, and this is likely to be related to the sort of action which people perceive they can take against the differing threats. For example, moving house or changing jobs will solve the threat to an individual's safety from a particular chemical plant, but will do nothing for the threat to societal safety from that same plant.

Despite these differences there appear to be common factors which people use to make assessments of danger and to apply a label to a situation indicating that something must be done about it. What varies between types of hazard and between responses to different questions is the weighting given to the different factors.

The research has used two basic approaches. Either to ask people directly what they think about hazards and how they react to them, or to consider actual behaviour in respect of different hazards. The first is called expressed preference research, the second revealed preference.

One clear result of expressed preference research is that people use a more sophisticated assessment process in judging risk decisions than just considering probability of harm. They also consider a wide range of other factors, which can be grouped under the following headings:

1 Whether the victim has a real choice to enter the danger or not, or to leave it once exposed.
2 Whether the potential for harm in the situation is under the control of the potential victim or another person, or outside any human control.
3 The foreseeability of the danger.
4 The vividness, and severity of the consequences.

4.6.3.1 Choice to enter and leave danger

Those who choose to engage voluntarily in activities like skiing which they know to be dangerous seem to be considered to know what they are doing, to realise the nature of the hazards and to have accepted their own responsibility to control them. The problem of accidents or disease is then seen as their affair, and they are presumed to be in control of the hazard. On the other hand if there is no choice about exposure to the danger, e.g. in having a nuclear plant built near your village, far higher demands on the level of safety are made. The situation is, however, seldom clear cut. Can, for example, the choice of a person to take a job on a construction site in an area of high unemployment be called a voluntary acceptance of risks associated with that job? Hazards frequently come as part of a package with other costs and benefits. In the early years of the industrial revolution workers were deemed to have accepted voluntarily the hazards of the job that they accepted. Therefore they were deemed liable for their own accidents. Now both society's view and the law have changed. The employee is not considered to accept occupational hazards voluntarily, unless there is talk of some gross deviation from normal carefulness.

The demand for increased safety levels is also stronger if the risks and benefits are not equitably shared and one group profits from the risk exposure of another.

There is some evidence that dangerous activities which are voluntarily chosen are positively valued partly because of their finite element of danger. Mountaineers choose to attempt climbs of increasing difficulty as their skill increases, finding the old ones tame. There is an element here of testing the degree of control which one has over a situation to check that it is real. The element of apparent loss of control is one of the attractions of fairground rides such as the 'wall of death'. But the fascination seems to go further than this. Greater danger, such as in war or time of disaster, is associated in the minds of survivors with greater group friendliness, shared feelings, sense of purpose and competence which makes that danger in retrospect positively valued, or at least willingly accepted.

4.6.3.2 Controllability

The largest element here seems to be the feeling of personal control. Those who believe themselves knowledgeable about and in control of a dangerous situation, even where the magnitude of the consequences is potentially great, show little fear or concern about it. This also applies to attitudes towards the safety of others. Thus construction site supervisors may consider[32] that the site hazards are under the control of skilled craftsmen and not personally concern themselves with them, even when they see that that control is not being fully exercised. Similarly workers in a plant are much less concerned about the hazards from it than those who live nearby but do not work there.

If the assessment of personal control is such an important factor in evaluating hazards it is very important that the assessment is accurate, and that people do not believe they are in control when they are not. But there is ample proof that people can have illusions of great control where none or less exists. Svenson[33] quotes a number of examples from the field of driving. For example between 75 and 90% of drivers believe themselves to be better than average when it comes to driving safely; only 50% can be right. Similarly 88% of trainees in cardiopulmonary resuscitation felt confident after an interval of several months to perform it, while only 1% actually performed adequately. Experts always tend to be overconfident in their expertise. This is particularly dangerous when specific knowledge, for example of a theoretical nature, about a process is taken to mean control over the whole activity involving that process. This can be a serious source of overconfidence in skilled personnel such as research chemists or toolroom personnel, most of whose accidents in fact come from the everyday hazards of the machinery or the laboratory which have little to do with their speciality.

When people have no personal control they may place their trust in others to keep the situation safe. Again it is a question of whether the assessment is accurate and whether the trust is justifiably placed. The work of Vlek & Stallen[30] suggests that one of the clusters of beliefs characterising those who oppose large nuclear, transport and chemical plant developments is personal insecurity and lack of trust in those controlling the technology. The situation is made worse by the spectacle of experts disagreeing violently with each other about the safety issues of the developments in question.

In the field of health promotion the concept of control has also been shown to be important. One of the main thresholds to be crossed before people will act to change their own behaviour is to admit that they personally are susceptible to the health threat, for example, from smoking, alcohol, drugs, or heart disease, i.e. that they have lost control. The other side of this coin is the need to believe in the efficacy of the preventive action before it will be adopted. This means believing that the proposed action would restore the lost control; that giving up smoking would reduce the risk of cancer and heart disease, that wearing the protective earmuffs would reduce the hearing loss and so on. The opportunity to prove for oneself the effectiveness of protective devices is therefore important in persuading people to wear them.

4.6.3.3 Foreseeability

Foreseeability is a word familiar from the case law of the English legal system relating to health and safety. It has been defined by judges with reference to what the 'reasonable man' would expect to happen given access to the current state of knowledge at the time of making a decision. It is used to draw a dividing line between situations in which people should have taken action to prevent an accident, and those for which it is not reasonable to hold them liable.

At an individual level this also affects the assessment of risk with hazard detection being limited by what is foreseeable or foreseen. But if people cannot foresee exactly what may happen, but suspect that it may still go wrong, they will be afraid. If this feeling goes hand in hand with the belief that there could be severe consequences and that the person is powerless to do anything, the reaction may be extreme. Evidence that a new and unknown technology like genetic engineering is not as much under control as previously thought would therefore have a profound effect on people's beliefs, shifting them rapidly from indifference to strong opposition. This is approximately the effect which Chernobyl had on nuclear power.

4.6.3.4 Vividness, dreadfulness and severity

The most recent accident or tragedy weighs heavily in the minds of people when they are asked about priorities for prevention, but this may fade rapidly. On a more permanent basis, people have reasonably consistent ratings of how nasty particular types of injury or disease are[34]. For example cancer is greatly feared, an eye is worth more than a leg and some injuries such as quadriplegia and brain damage are consistently rated as worse than death.

An important element in memorability is 'kill size', the number of people who either actually do, or potentially could, get killed in an incident.

4.6.3.5 Conclusion

With such a complex of factors determining the reaction of both individuals and society to risk it is not surprising that no simple scale such as Fatal Accident Frequency Rate can capture its essence. Managers and planners may wish to reduce decision making to a tidy consideration of probability times cost of harm (usually deaths). They may even wish to label as irrational any opposition to this definition of risk. But this is no more than one powerful group putting an emotive label on something they seek to oppose. A better approach is to treat the factors for what they are, namely the basic elements which must be influenced if we wish to change behaviour. If you want someone to use a safety device, you must convince them that the danger is foreseeable, unpleasant and avoidable, that the safety device is effective and that they can choose how to use it.

4.6.4 Assessment of probability

For a normal person probability is not a concept that comes naturally. This is an idea that will be readily accepted by anyone who has tried to learn the fundamentals of statistics. Most normal people have little need for accurate probability judgements and little practice in making them. Individuals normally only rate whether they think things will remain under control. Despite this, it is surprising how good the correlation is between measured probability of particular types of accident and subjective assessments by the general population. The major bias is that the subjective scale is compressed and foreshortened in relation to the objective, typically spanning only three orders of magnitude instead of six. Very rare risks are treated as non-existent, slightly less rare risks may be overestimated and common ones underestimated. Some hazards are raised in the order of probability; typically those which receive media coverage.

The framing of questions about probability and of statements about risk can strongly affect people's responses to them. It is more effective as an argument to get people to be vaccinated to tell them that a vaccine offers total protection against one strain of disease that accounts for half of a given sickness, than to tell them that the vaccine offers 50% protection against the sickness; the presence of the word 'total' in the message gives the illusion of certainty. Framing information about road accidents in terms of the probability of accidents over a lifetime (probability of death c. 0.01 – and of disabling injury c. 0.33) is much more effective in getting people to wear seat belts than quoting the probability over one trip (probability of death c. 1 in 3.5×10^{-6}, and of disabling injury c. 1 in 10^{-5}).

4.6.5 Responsibility for action

Even if people can see a danger and appreciate the need for action, they may not act because they think they cannot or should not. This may be because it is seen as someone else's job or responsibility. This attitude is found among supervisors who are not willing to tell skilled workers how to avoid risks in their job[32]. Social pressures determining what is or is not acceptable behaviour may discourage people from warning others because of the fear of being told to mind one's own business, or of being thought to be interfering.

The major factor at this stage will be the way in which people view the courses of action open to themselves and others to influence the danger. Again the crucial role of the mental models of cause and effect are clear. If I believe as a supervisor that accidents to my staff are caused by their own carelessness and lack of attention to rules I will only think in terms of selection, training and discipline as actions. If I believe that the machine design is such that nobody can be expected to concentrate 100% of the time to avoid injury, I shall give attention to redesign or guarding as well. Biases in the way people allocate responsibility for accidents or prevention are therefore of vital importance and should be the subject of training. Such biases can be summarised as follows[1]:

1 When people are looking at their own future behaviour they think that they can exercise more control than is usually the case. Hence they accept great (even too great) responsibility to act to control the situation and to prevent future accidents.
2 When people personally suffer an accident, they are inclined to attribute it too much to the force of external circumstances rather than to personal responsibility.
3 When people observe others' behaviour they grossly underestimate the effect that the situation has on determining it; hence they overestimate the control that others have over what they do, and blame them unfairly for their accidents. This can lead to a reluctance to intervene in situations to warn, instruct or help people.

These biases arise when there is some ambiguity in a situation which allows for more than one interpretation. Such occasions are most frequent in rapidly changing situations, and when people are trying with hindsight to reconstruct an accident of which they may have been a witness or about which they have merely heard reports. Putting the three biases together goes some way towards explaining the inactivity in accident prevention in a number of situations. Designers must consider hazards to others (the users). They tend to overestimate users' ability to look after themselves, and so underestimate the need to build in safeguards. Supervisors place the onus for avoiding accidents on the victims and not on themselves; while their bosses think that it is the supervisors' job and so shuffle off their own responsibility for the climate of rules and priorities which they create. If managers and designers sit round a table with workers from the shop floor who are talking about their own accidents (and so are subject to the first two biases), a great measure of agreement is possible. Both sides will agree that the main onus lies on the worker to prevent accidents. However, this agreement does not lead to any action, as both sides will also tend to believe that everything is under control and nothing needs to be done.

4.6.6 Action plans

The major problem under this heading is whether the necessary actions have been learned and are available when needed. The latter point is particularly crucial in emergencies, such as evacuation, plant shut-down and first aid treatment. Considerable investment in refresher training, sometimes on simulators, is needed to keep rarely used skills available enough to be deployed accurately when they are wanted.

4.7 Change

Underlying the above discussion of human characteristics, limitations and differences has been the notion of change. The human system is in dynamic equilibrium with its physical, social and cultural environment, constantly adjusting itself to changes in that environment while it pursues its goals and objectives. The environment changes as society and technology change and as family responsibility and job type and location change. At the same time the individual changes, grows up, matures and grows old, learns new skills, forgets old knowledge and acquires new goals. Change is intrinsic in the human condition. It is a mistake to think of solutions to problems in health and safety in terms of changing a person from one stable state to another. The problem is better seen as one of prodding and guiding behaviour along one path out of the many possible ones and trying to stabilise it in a new path which will maintain its essential characteristics while it carries on adapting to other changes in the world. An appropriate analogy for change is steering a sailing ship on a turbulent ocean rather than the action of picking up a piece of plasticine, remoulding it and setting it down again.

In the following sections the various processes of change are briefly presented. First the 'natural' processes of growing up and change are sketched, which lead to individual differences in personality, motivation, knowledge and skill. Then the process of learning is outlined, which must form the basis for the systematic acquisition of experience in training courses.

4.7.1 Growth, change and individual differences

Individuals differ because of three influences: genetic or inherited characteristics, learned or environmental characteristics, and situational factors operating at a particular instant.

4.7.1.1 Genetic factors

Many of the physical characteristics of individuals are strongly influenced by genetic factors, for example hair and eye colour, height, physical build and body dimension, although the last three are also influenced by environmental factors such as nutrition. When it comes to factors such as intelligence or personality there is far greater argument about the importance of genetic factors in determining the final measured characteristic in an individual. This debate is important wherever we are trying to change behaviour through education or by changing social or company policy, since what is genetically determined is likely to be more or less unchangeable.

4.7.1.2 Environmental or learned characteristics

Environment forms the most important influence on the personality and social characteristics of individuals. This category is often sub-divided according to which stage of development or aspect of the environment is responsible for the influence.

The newspaper headlines of recent years are eloquent witnesses to some of the more dramatic effects occurring between conception and birth which can produce individual differences. Drugs such as thalidomide have massive deforming effects, as do diseases such as German measles contracted during critical phases of pregnancy. Alcohol and cigarette consumption in pregnancy have also been shown to have less severe but widespread effects on the foetus.

The opportunities provided by the family, and the goals and the motivations which are learned because of the behaviour which is rewarded during childhood have a great effect on the attitudes, skills and abilities shown in maturity. During later childhood and adolescence school, teachers and peer groups take over from the immediate family as the major influence on the development of personality and attitudes. The rate at which personality and attitudes change slows down once the age of 20 is reached, but the working environments and the social groups within which time is spent still have a formative and changing influence. Individuals tend to seek out social groups which suit their personality and attitudes, but when there is a mismatch the individual will be changed by the group to a greater or lesser extent depending upon the importance that the individual attaches to acceptance by the group.

Some of the factors which have been mentioned are often drawn together and labelled as cultural factors if they are influences which are shared by a defined group of people, whether in a particular country, region, social class, age, or occupational group. Thus certain attitudes towards risk taking may be shared by members of one adolescent culture, which are markedly different from those of other age groups.

4.7.1.3 Situational factors

Besides the innate and learned factors which an individual carries around all the time, behaviour at any one moment will be influenced by situational factors. For example, people slow down when a police car drives past, get excited when their team scores a goal or conceal their political views to avoid arguing with their prospective father-in-law. Situational factors can sometimes make people behave in ways which are untypical of them, or of which they disapprove.

4.7.1.4 Personality and attitudes

Since each individual will have been subject to a unique mixture of all of the factors set out above, the result is that no two individuals will be entirely alike in the combination of characteristics which make up their behaviour. No two people will perceive the world in quite the same way. No two individuals will react in quite the same way to the same circumstances confronting them. To predict with certainty how any one individual will behave in a particular set of circumstances would require a complete knowledge of all the factors which had gone to make up that person. However, the position is not entirely hopeless since there is enough common ground in individual responses to most circumstances to make predictions worthwhile. That common ground within one person is labelled personality; where it is common ground between people in a group we label it norms or group attitudes.

The study of personality is an area of psychology which has spawned many parallel and conflicting theories. One style of theory tries to explain where personality comes from and classifies people into 'types' or groups based on differences in personality development; other theories merely classify the end result and measure existing differences (trait theories).

Freudian psychology defines personality as developing through phases of coming to terms with the conflicts between the instinctual and social parts of a person and in learning to express and control sexual drives. It classifies people according to the degree to which their full development and resolution of these conflicts is incomplete. The insights of Freudian analysis have been enormously influential in many fields, but apart from enriching language and the arts, their importance lies mainly in understanding mental pathology.

Other developmental theories stress social influences more (Laing), or concentrate on the cognitive development of the means of making sense of, and classifying events and its influence on personality (Kelly).

Cattell's trait theory is taken as an example of a descriptive theory. From extensive research based upon the responses to questionnaires on their beliefs and preferences by many thousands of individuals, Cattell has produced a list of 16 personality factors (see *Table 4.2*). The factors are envisaged as 16 dimensions on which an individual's position can be plotted to produce a profile which describes that unique individual. Since someone can score from 1 to 10 on each scale, these 16 scales provide 10^{16} unique character combinations or personalities, which is more than the total number of human beings who have ever walked the earth since Homo sapiens evolved.

Table 4.2. Cattell[35] personality factors

1 Reserved, detached, critical	Outgoing, warm-hearted
2 Less intelligent, concrete thinking	More intelligent, abstract thinking
3 Affected by feelings, easily upset	Emotionally stable, faces reality
4 Humble, mild, accommodating	Assertive, aggressive, stubborn
5 Sober, prudent, serious	Happy-go-lucky, impulsive, lively
6 Expedient, disregards rules	Conscientious, persevering
7 Shy, restrained, timid	Venturesome, socially bold
8 Tough-minded, self-reliant	Tender-minded, clinging
9 Trusting, adaptable	Suspicious, self-opinionated
10 Practical, careful	Imaginative
11 Forthright, natural	Shrewd, calculating
12 Self-assured, confident	Apprehensive, self-reproaching
13 Conservative	Experimenting, liberal
14 Group-dependent	Self-sufficient
15 Undisciplined, in self-conflict	Controlled, socially precise
16 Relaxed, tranquil	Tense, frustrated

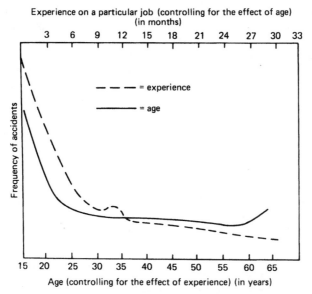

Figure 4.8 Distribution of accidents by age and experience. (Derived from Hale and Hale[10])

4.7.1.5 Individual differences in accident susceptibility

Research into individual differences in accident rate (accident proneness) has a long and complex history littered with mistaken conclusions from invalid methodology and poor experimentation[1,10]. It set out to discover whether there were stable differences in accident susceptibility when individuals were subjected to equal hazards (both in type and length of exposure). It was established very early in the research that both age and experience were correlated with differences in accident susceptibility (see *Figure 4.8*). The exact shape of the graphs will vary from job to job.

Job-related experience appears to be that most relevant to accident rate (Powell et al.[36]) although the effects of number of years in industry and of number of hours worked on a specific task (where the job involves a number of tasks) can also be demonstrated.

The relationship of physical and anthropometric differences to accident susceptibility has also been shown in many specific tasks. For example, colour blindness can be a danger where hazard perception depends on colour discrimination; extremes of height, reach and slimness of arms, wrists or fingers can result in individuals being able to reach into danger areas around or through guards; susceptibility to epilepsy, bronchitis or eczema can be problems on jobs involving moving machinery, dust and oils respectively; etc. Research on sex and ethnic differences in accident liability often shows apparent differences in gross accident rates, but these almost always turn out on closer examination to be differences in risk exposure (i.e. immigrant workers and men tend to be found more often in the dirtier and more dangerous jobs).

Research on the relationship of other factors to accident susceptibility has produced few clear-cut results; personality factors, intelligence coordination and attention skills and many other characteristics have been studied but the correlations produced have usually been low and have been specific to the job or task studied. Accident proneness as an explanation for accidents or a basis for a safety policy is therefore very unprofitable and only helps to reinforce a blame culture instead of a problem-solving one.

4.7.1.6 Attitude

Attitude is sometimes defined as 'a tendency to behave in a particular way in a certain situation'. Underlying this definition is one of the thorniest problems in psychology, the consistency between what people say they believe or will do and what they actually do. As with personality many theories abound in this area. In their theory Fishbein and Ajzen[37] define:

Attitude.   Attraction to or repulsion from an object, person or situation. Evaluation, e.g. liking children, favouring trades unions etc.
Belief.   Information about an object, person or situation (true or false) linking an attribute to it, e.g. that guards are a hindrance.
Behavioural intention.   A person's belief about what he will do if a given situation arises in the future, e.g. that he will use his safety belt when driving on a motorway.
Behaviour.   Actual overt action, e.g. telling the interviewer that you will wear your seat belt; actually doing so.

All these are linked as shown in *Figure 4.9.*
Thus, for example, someone may believe that breathing apparatus is uncomfortable and dislike it (which may feedback to beliefs by making that person hypercritical of the comfort of any new apparatus) to the extent that the person develops resistance to wearing it, but knowing that

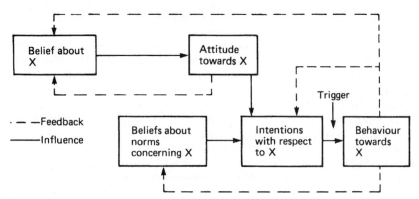

Figure 4.9 System diagram showing links between attitudes and behaviour

it is a company rule (norm) will hurriedly put it on (behaviour) when the safety adviser walks by (trigger). If this happens many times he may find it is not so bad after all and there will be feedback which will change the beliefs.

4.7.2 Stimuli for change and resistance to change

A stimulus for the individual (or indeed for the organisation) to change will be the perception that his or her adaptation to the environment is no longer as close as desired. If the failure in adaptation is not perceived there will be no acceptance by the individual that change is needed. In that case attempts to impose change will be met with the sort of resistance which is characterised by the remark 'We have done it this way for 50 years and it's been OK; why change now?' This conservatism in attitudes and beliefs seems to become more marked with advancing age, perhaps because there is more past experience to call on for support in rejecting the need for change.

Social groups are frequently bastions against change. If a number of people can be found to share the view that things are all right as they are, they reinforce each other's view of the world and unite to resist change.

Those wishing to promote change have the task of convincing individuals, groups and organisations that change is needed because the old adaptation is no longer appropriate. This process is least difficult when something dramatic such as an accident demonstrates that old methods were not safe, or there is a new law, a new machine, a new boss or a new job, all of which self-evidently require change.

It is the slow changes in environment and individuals which often go undetected, and so unresponded to. Examples are changes in technology, affluence and life expectancy or in attitudes to work which occur as the result of growing up or of growing old. If these changes are to be brought to the notice of individuals or organisations it is often necessary to dramatise them in order to get through the conservatism or blinkered vision which is failing to see them.

4.7.3 Methods of change

Change can be brought about by:

1 Changing the physical or social environment of the individual so that the old behaviour does not bring the rewards it did, or new behaviour brings greater rewards.
2 Providing new information to an individual which shows that the existing adaptation is not as close as the individual thought it was, or as it could be.
3 Changing the goals and objectives which an individual is seeking.

4.7.3.1 Changing the environment

Under this heading come modifications to physical work design and layout, new machinery and work methods, changes in allocation of jobs to people, introduction of new incentive schemes so that different behaviour is rewarded and punished (e.g. payment schemes, promotion systems, direct safety incentives or simply what behaviour elicits praise from the boss), changes to safety rules, company policy or legal standards.

In all cases the change must be brought to the notice of individuals, and conformity to the new rules or situation must be appropriately rewarded and persistent non-conformity punished. Provided that the person, group or organisation introducing the change is perceived to have sufficient authority (or legitimate right) to make the change, or is sufficiently powerful in wielding reward and punishment, or has great enough charisma or expertise, the change will be accepted and adapted to.

4.7.3.2 Giving information on maladaptation

Under this heading fall the provision of information about danger, provision of training on more effective ways of conforming with any of the changes outlined in section 4.7.3.1 above, or pointing out the way in which behaviour in one area conflicts with strongly held beliefs in another area. Another approach is to improve communication skills so that people are more open to influence (e.g. sensitivity training). The success of all these endeavours will depend upon the credibility of the source of information and the ability of that person or organisation to organise and put over information.

4.7.3.3 Changing long-term goals and objectives

Under this heading come education, media and advertising campaigns to build and change 'images' and also long-term changes in law and culture, etc. These methods of change are usually very long-term, and are often poorly understood. They operate by training people to look at, question, and so develop their own goals. They also expose people to different opportunities and chances for achievement, and present them with examples of what are labelled 'acceptable' and 'unacceptable' behaviour for people to copy.

4.7.4 Learning and training

This section deals with the psychological principles of learning. A number of authors have classified learning processes into a hierarchy of levels, building from the simplest stimulus response learning to the most complex processes of research and problem solving. Three levels of learning are particularly relevant:

1 Stimulus response learning (Skill).
2 Concept and rule learning (Rule).
3 Problem solving (Knowledge).

4.7.4.1 Stimulus response learning

This is the building block which is particularly important in all 'habits' and repeated invariant sequences of behaviour. The sequences are built up by a process called conditioning. This has three essential elements:

1 Evoking the correct response when the stimulus is presented; by trial and error, explanation or demonstration.
2 Rewarding correct responses: the rewards can include praise for correct performance, administering punishment for incorrect performance and the use of monetary rewards. It is also possible to play upon the motives of achievement and interest by feeding back information of how well the task is being learnt or how near to the objective the learner has reached.
3 Practice of the correct sequence which establishes the response more and more firmly until it occurs without conscious effort.

Once sequences of action of this sort have become established it is extremely difficult to add to them or subtract from them. It is therefore important that in safety all the necessary steps are built into the sequence at the learning stage. For example, if the response of donning protective goggles is built into the sequence of setting up and starting an abrasive wheel, it will become an automatic part of that habit. If some responses are not built in at the time of first learning, then either the behaviour sequence must be broken down and relearned with them in place, or, as a poor second best, individuals must be trained consciously to break into the chain at the appropriate point in order to carry out the missing response action.

The sequences of actions will become less automatic if they are not used regularly. Behaviour in response to infrequent emergencies will therefore not be available unless it is practised in the interim.

4.7.4.2 Concept and rule learning

A fundamental human characteristic is the tendency to think about and classify objects, experiences or ideas into mental categories or 'concepts', which share things in common.

Concepts are built up from experience and each person's set of concepts will then differ slightly. Thus, for example, the concept 'dangerous' to one individual may contain hundreds of items which include any situation which is new or strange (a person we might label 'nervous'); another's might contain very few items and leave out some which should be there, such as noisy discotheques or bottles of weed killer kept in the larder.

Acquiring concepts depends on amassing examples which have elements in common until the individual arrives at a tentative definition of the boundaries of the group of objects or ideas. This process can be abbreviated by defining the concept for the person. It must then be consolidated by providing examples which fall inside it, and outside it, gradually reducing the difference between these positive and negative examples until the boundaries become clearly defined. People continue even then to test out and modify the boundaries of the concept.

Concept learning is appropriate whenever someone is expected to recognise a new stimulus as belonging to a particular category, so that they can respond to it even though they have never met it before. It is also necessary whenever someone is expected to follow a rule. For example the rule 'all hazards must be reported to a responsible person' requires that individuals should learn a correct concept of what is a hazard, and a correct concept of who are responsible persons, as well as an adequate idea of what reported means (a casual mention, a formal verbal report, or a formal written statement).

4.7.4.3 Problem solving

Where someone is faced with a situation which they have never met before, and cannot clearly place it into an existing concept category, they are faced with the need to produce a solution new to them. Problem solving is a creative process which relies on the basic building blocks described above and consists of re-ordering them and re-interpreting them.

Learning to solve problems can be aided by teaching the steps in systematic problem solving:

1 To recognise the problem area and define it in broad terms.
2 To explore all the possibilities for solving the problem.
3 To analyse all the facts available to determine whether or not subsequent problems will follow from the solution suggested.
4 To choose the best possible solution and implement a plan of action for introducing it.

Some techniques for creative thinking (step 2) can also be taught, e.g. brainstorming. A problem is presented to a group of people who are encouraged to throw out ideas, sparked off by each other, no matter how wild. These are recorded by the group leader on a board for all to see. It is important that all ideas be put forward without either interruption or criticism because cross-fertilisation and the building up on ideas produced by others will generate many new ideas. After a suitable period, the ideas noted can be discussed in open forum as part of the evaluation stage to obtain the best solution to the problem from the large pool of ideas so formed.

4.7.5 Communication

Learning and attitude change depend on the motivation to change and on good communication of the message about what change is needed.

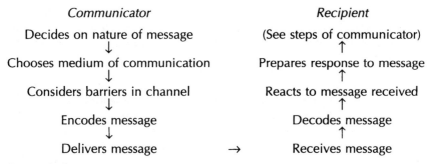

Figure 4.10 Steps in communication

Earlier sections have dealt in passing with aspects of communication: warnings, information about hazards and prevention measures, check-lists and training courses. In this section the basic rules for the process of communication are covered.

Communication is the process whereby one person makes his ideas, feelings and knowledge known to others and learns in exchange about theirs. It is therefore a two-way process which depends crucially on both clear sending and receipt of the message (Figure 4.10).

The first essential is to know with whom you are communicating (directors, line managers, accountants, the workforce, the factory inspec-torate). Then the precise objectives of the message must be planned (what change is wanted? what are the precise obstacles? must the message succeed with more than one different group?). The message must be coded in terms appropriate for the audience. It must latch onto their way of thinking, priorities and language. It must not use jargon they do not know. To do all of this it is necessary to think about the subject of the communication from the point of view of the receiver and to plan the message so that it leads from that point, covers the disadvantages of the change from that viewpoint and how they will be overcome and ends by spelling out the advantages of the change.

On the basis of this planning the medium can be chosen:

1 *Face to face communication* has the advantage that it allows feedback and adjustment of the message based on the response. It is also friendlier and less formal. However it is less easy to control because it is interactive. It is important to remember that it consists of two elements, verbal and non-verbal. The words seem dominant and must indeed be chosen appropriately and put over clearly, but the non-verbal clues can either reinforce or destroy their effect. The tone of voice can indicate boredom, the stance friendliness, hand move-ments nervousness. The very different effect of messages over the telephone and face-to-face demonstrate this point. It is excellent practice for communicators to listen to themselves on tape and watch themselves on video to see and correct these elements of their style.

2 *Written communication* allows for much more complex messages to be sent and understood because they can be reread and carefully weighed. It also forms a permanent record for future reference, both for and against the sender.

3 *Visual communication* allows for very rapid transmission of the relations between things in one glance. It can therefore have great power and emotional impact, but is less easily controllable. There is just as much a language of pictures which must be learned by both parties to the communication, as there is a language of words.

4.8 Methods of change and control

Finally two topics will be briefly reviewed which have been behind much of the discussion in this chapter:

● design as an influence on behaviour
● the response to safety rules.

4.8.1 Design

Ergonomists have been saying for a generation that design is the biggest influence on use. The operator or user has too often in the past been saddled with the impossible task of recovering the inadequacies of the designer. Machines which operate in ways which do not fit expectations, routines which are easily confusible, and tasks which have been left to the operator only because technology cannot yet take them over, are all examples of accidents planted like time-bombs in the system. Users are to be congratulated that they manage for so much of the time to operate safely despite them. The increase in feedback from users to designers and the heavier emphasis on product liability which have occurred in the last decade (e.g. in the EC Product Safety Directive) are welcome signs that this tolerance is reaching its limit.

Designers should also be humble enough to realise that their designs are not eternal and that operators have a need to modify or adjust their workplace and not to have to work within rigid constraints which are not perfectly attuned to operating conditions. Therefore, enough information and training must be provided to the operators to enable them to recognise the possibilities for modification and prevent them from falling into unsuspected traps; in addition, predictable modifications which will lead to danger (like removing guards or defeating safety interlocks) should be made as difficult as possible.

Designers should not think of people as automata. There are differences both between and within individuals. People will never be as consistent in their response as hardware components. Therefore designs must be error tolerant and make error recovery as easy as possible. Designers must predict foreseeable misuse and error and cater for it. This is now a principle enshrined in European machinery standards[38]. Nor should designers respond to human error with an unthinking push to automate the individual out of the system as much as possible. That is a recipe for creating residual monitoring tasks which are boring and unsatisfying. It will also result in the loss of skill and insight to such an extent that the operator cannot intervene effectively when the hardware fails.

Finally, designers should not have unrealistically high hopes of their hardware solutions. People will always adapt to system changes by altering their behaviour, sometimes trading off increased safety margins against other gains; e.g. straightening out roads with dangerous bends results in an increase in traffic speed; more reliable hardware results in less spontaneous checks of its functioning. This behaviour, sometimes called risk compensation[1], should be anticipated by the designer, who should design against such trade-offs.

4.8.2 Safety rules

We should be suspicious of anyone who claims that safety is merely a matter of laying down and enforcing rules. It can never do any harm to define clearly and as exhaustively as possible how the system should operate to overcome all known hazards. But enforcement of such detailed rules is difficult to achieve. This approach takes away all individual freedom and control over the work. It will only work where danger is very evident and it can be guaranteed that application of the rules will always result in safety. Even then it will work only with difficulty if following the rules is also not the easiest and most obvious way of doing the job.

The following extract from a study of rules[39] is typical: 50 railway workers were asked about safety rules governing work on and near railway tracks;

- 80% considered that the rules were mainly concerned with pinning blame.
- 79% thought there were too many rules (12% too few).
- 77% found the rules conflicting.
- 95% thought that work could not be finished on time if the rules were all followed.
- 85% found it hard to find what they wanted in rule book.
- 70% found the rules too complex and hard to read.
- 71% thought there was too little motivation to follow rules.
- Not one could remember ever having referred to the rules in a practical work situation.

Rules are subject to exceptions and to erosion. Safety manuals and safety laws tend to be full of complex specifications with many 'if . . ., then . . .' clauses which are perfect if followed, but which are too complex to remember. Execution of all the checks to see which sub-clause applies in any one case would often take too long in practice. Such rule books only serve to assuage the conscience of the rule makers. After an accident they can establish exactly who should have done what and so who was to blame. The existence of such a complex edifice of rules is a signal that the system is inwardly sick and in urgent need of redesign to incorporate behavioural rules into either training or hardware design. Ideally, design should precipitate the right action, and articulated written rules are only necessary where the way someone would expect to have to operate in a given situation is not in fact correct.

This conflict between establishing rules and leaving the flexibility to cope with exceptions and with changes can be seen at all levels in safety. It is reflected in the arguments about rigid central specification in laws and standards in contrast with enabling frameworks with objectives and freedom for each company to comply in the way it wishes. It can be seen at the level of the company where operating managers are keen to reduce problems as fast as possible to fixed rules in order to be able to get on with production. Safety departments have a task here to act as the protagonists of continual revolution in the firm. Safety rules need to be written with the involvement of those who must follow them. They also need to be updated at regular intervals using the critical experience of those same people.

References

1 Hale, A. R. and Glendon, A. I., *Individual Behaviour in the Control of Danger*, Elsevier, Amsterdam (1987)
2 Hale, A. R. and Hale, M., Accidents in perspective, *Occupational Psychology*, **44**, 115–121 (1970)
3 Taylor, F. W., *Principles of Scientific Management*, Harper & Row, New York (1911)
4 Mayo, Elton, *The Social Problems of an Industrial Civilisation*, Routledge & Kegan Paul Ltd, London (1952)
5 Maslow, A. H., *Motivation and Personality.* Harper, New York (1954)
6 Atkinson, J. W., Motivational determinants of risk-taking behaviour, *Psychological Review*, **64**, 359–372 (1957)
7 McClelland, D., Risk-taking in children with high and low need for achievement. In Atkinson, J. W. (Ed.), *Motives in Fantasy, Action and Society*, van Nostrand (1958)
8 Porter, L. W., Lawler, E. E. and Hackman, J. R., *Behaviour in Organisations*, McGraw-Hill, Kogushawa, Tokyo (1975)
9 Adler, A., The psychology of repeated accidents in industry, *American Journal of Psychiatry*, **98**, 99–101 (1941)
10 Hale, A. R. and Hale, M., A review of industrial accident research literature, Committee on Safety and Health at Work, *Research Paper*, HMSO, London (1972)
11 Wagenaar, W. A. and Groeneweg, J., Accidents at sea: multiple causes and impossible consequences, *International Journal of Man-machine Studies*, **27**, 1–90 (1987)
12 Binet, A. and Simon, T., The development of intelligence in the child, *Année Psychologique*, **14**, 1–90 (1908)
13 Report of the Commissioners appointed to collect information in the manufacturing districts relative to the employment of children in factories. *Parliamentary papers* – three reports, 1833/34
14 Waterhouse, J. M., Minors, D. S. and Scott, A. R., Circadian rhythms, intercontinental travel and shiftwork. In Ward Gardiner, A. (Ed.), *Current Approaches to Occupational Health*, **3**, Wright, Bristol (1987)
15 Mackay, C. and Cox, T., *A Transactional Approach to Occupational Stress*, Department of Psychology, University of Nottingham (1976)
16 Cox, T., *Stress*, Macmillan Press, London (1978)
17 Reason, J. T., A framework for classifying errors. In Rasmussen, J., Leplat, J. and Duncan, K. (Eds.), *New Technology and Human Error*, Wiley, New York (1986)

18 Rasmussen, J., What can be learned from human error reports. In Duncan, K., Gruneberg, M.M. and Wallis, D.J. (Eds.), *Changes in Working Life*, Wiley, Chichester (1980)

19 Hale, A. R. and Pérusse, M., Perceptions of danger – a prerequisite to safe decisions, *Proceedings of the Institution of Chemical Engineers*, Rugby (1978)

20 Surry, J., *Industrial Accident Research*, Department of Industrial Engineering, University of Toronto (1969)

21 Tong, D., The application of behavioural research to improve fire safety, *Proc. Ann. Conf. Aston Health and Safety Society*, Birmingham (1983)

22 *Procedures Guide for Probabilistic Risk Assessment*, U.S. Nuclear Regulatory Commission Report NUREG/CR 2000 (1980)

23 Swain, A. D. and Guttman, H. E., Handbook of human reliability analysis with emphasis on nuclear power applications, *US Nuclear Regulatory Commission Report NUREG/CR 1278*, Sandia Laboratories, Albuquerque, New Mexico (1983)

24 Tettelaar, H. C., de Vries, K. L. M. and Phaf, R. H., *General Risk Assessment Procedure (GRASP)*: Incorporating Human Error into Risk Analysis. Report R-88/28, Centre for Safety Research, University of Leiden (1988)

25 Feggetter, A. J., A method for investigating human factors aspects of aircraft accidents and incidents, *Ergonomics*, **11**, 1065–1075 (1982)

26 Kirwan, B., Human reliability assessment. In Wilson, J.R. and Corlett, N. (Eds.), *Evaluating Human Work: an Ergonomics Methodology*, Taylor and Francis, London (1989)

27 Royal Society, *Risk Assessment*; a Study Group Report, London (1983)

28 Lowrance, W., *Of Acceptable Risk: Science and Determination of Safety*, W. Kaufman, Los Altos, California (1976)

29 Health and Safety Executive, *The Tolerability of Risk from Nuclear Power Stations*, HSE Books, Sudbury (1988)

30 Vlek, C. and Stallen, P-J., Judging risks and benefits in the small and in the large, *Organisational Behaviour and Human Performance*, **28**, 235–271 (1981)

31 Starr, C., Social benefit versus technological risk, *Science*, **16**, 1232–1238 (1969)

32 Abeytunga, P. K., *The role of the first line supervisor in construction safety: the potential for training*, PhD Thesis, University of Aston in Birmingham (1978)

33 Svenson, O., Risks of road transportation in a psychological perspective, *Accident Analysis and Prevention*, **10**, 267–280 (1978)

34 Green, C. H. and Brown, R. A., *The perception of, and attitudes towards, risk: Preliminary report: E2, Measures of safety*, Research Unit, School of Architecture, Duncan of Jordanstone College of Art, University of Dundee (1976)

35 Cattell, R. B., *The Scientific Analysis of Personality*, Penguin Books, London (1965)

36 Powell, P. I., Hale, M., Martin, P. and Simon, M., *2000 Accidents*, National Institute of Industrial Psychology, London (1971)

37 Fishbein, M. and Ajzen, I., *Belief, Attitude, Intention and Behaviour – an introduction to Theory and Research*, Addison-Wesley, Mass. (1975)

38 Centre Européen de Normalisation, *Safety of Machinery – Basic concepts – General principles for design. Part I: Basic terminology and methodology* (EN 292–1:1991). *Part II: Technical principles and specifications* (EN 292–2:1991). BS EN 292–1 and BS EN292–2 respectively, BSI, London (1991)

39 Hale, A. R., Safety rules OK? Possibilities and limitations in behavioural strategies, *Journal of Occupational Accidents*, **12**, 3–20 (1990)

Further reading

The primary text for further reading is:

Hale, A. R. and Glendon, A. I., *Individual Behaviour in the Control of Danger*, Elsevier, Amsterdam (1987). This chapter is in great part a summary of the material covered there in great detail. It also contains detailed references for still deeper reading.

Many of the following texts have an overlapping coverage of subject matter. The reader should therefore select from among them. The brief notes attached will guide that choice.

Canter, D., *Fires and Human Behaviour*, Wiley, Chichester (1980). A good review of work on the specific topic of reactions to fire.

Coleman, J. C., *Introductory Psychology*, Routledge & Kegan Paul, London (1977). Written with medical and nursing students in mind. The individual chapters are by different experts. Covers almost the full range of the subjects in these chapters.

Cohen, J. and Clark, J. H., *Medicine, Mind and Man*, W. H. Freeman & Co., Reading (1979). A parallel text to Coleman also written for students of health sciences.

Hoyos, C. G. and Zimolong, B., *Occupational Safety and Accident Prevention: Behavioural Strategies and Methods*, Elsevier, Amsterdam (1988). A parallel text to Hale and Glendon written somewhat more from the viewpoint of safety management.

Maier, N. R. F., *Psychology in Industrial Organisations*, 4th edn, Houghton Mifflin, Atlanta (1973). Written for industrial managers and supervisors. The examples are therefore good and relevant though with an American bias.

McCormick, E. J., *Human Factors Engineering*, 3rd edn, McGraw-Hill, New York (1970). A very thorough and complete source book for ergonomics and human factors in design.

Hale, A. R. and Hale, M., *A Review of Industrial Accident Research Literature*, Committee on Safety and Health at Work: Research Paper, HMSO, London (1972). A brief review of the literature up to 1972 on human factors in accident causation. Valuable source of further references.

Powell, P. I., Hale, M., Martin, P. and Simon, M., *2000 Accidents*, National Institute of Industrial Psychology, London (1971). Summary report of a four year field study of accident causes. Good overview of the priorities in the field.

Rasmussen, J., Duncan, K. and Leplat, J. (Eds.), *New Technology and Human Error*, Wiley, Chichester (1987). A very valuable book of readings of both theory and practice in human error assessment and control.

Reason, J., *Human Error*, Cambridge University Press (1989). An excellent book setting out the theories of a very influential researcher.

Stammers, R. B. and Patrick, J., *The Psychology of Training*, Methuen Essential Psychology E3, London (1975). Short text covering the main psychological approaches and insights into the subject.

Vroom, V. H. and Deci, E. L. (Eds.), *Management and Motivation*, Modern Management Readings, Penguin Books, London (1970). Fairly advanced text but an excellent summary of the many theories that exist in the area.

Readers wishing to keep up to date with research on this topic will find research and review articles in scientific journals such as *Safety Science* and *Applied Ergonomics*. Road traffic safety papers are to be found in the *Journal of Safety Research* and in *Accident Analysis and Prevention*.

5

Practical behavioural techniques

J. E. Channing

5.1 Introduction

Changing the behaviour of people at work to improve their safety is a feature of many aspects of legislation. The most obvious elements include the requirements for systems of work, the provision of information to employees, and requirements for training. These legal obligations have been in place for many years and are key elements of the Health and Safety at Work etc. Act 1974. In practice this approach alone has not eradicated accidents at work. This should be no surprise. The existence of written procedures and attendance at training courses are not particularly effective at gaining the correct behaviour from people preoccupied with their jobs and private thoughts when a hazard occurs.

From another perspective most businesses find that the majority of their accidents at work do not arise out of their specific technology or business activity. They arise mostly from everyday events such as slips, trips, falls and handling accidents. Yet trying to focus the workforce upon reducing these apparently trivial and unglamorous accidents is difficult. Paradoxically it is essential to attack all types of accidents. Focusing on some types of accident yet tolerating others is illogical. If a person slips the most likely outcome is a bruise, yet with a minor change in circumstance it could be fatal. An accident can be seen as a 'loss of control' and the consequences cannot be predicted with certainty. An approach which treats these 'everyday' accidents as unacceptable also promotes an attitude that prevents the obviously serious or catastrophic accidents. It also prevents the insidious chronic conditions which cause ill-health – such as musculoskeletal, skin or lung disorders – from occurring. This is a 'zero tolerance' approach to accidents.

Achieving a zero tolerance position requires a change of culture in the workplace and of the attitudes of people working in it. Terms like 'culture' and 'attitude' are easily understood in general terminology but quite difficult to develop into practical safety programmes. The relationship between attitude and behaviour is the subject of ongoing research but it can be argued that where a positive attitude toward safety exists correct safety behaviour occurs. A simple model linking attitude to behaviour is presented in *Figure 5.1*. This example examines the attitude

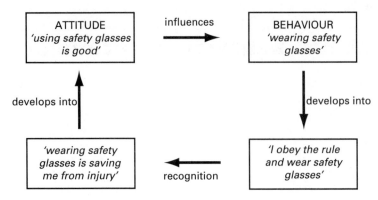

Figure 5.1 Attitude and behaviour: the wearing of eye protection

and behaviour towards the wearing of eye protection. People who normally do not need to wear spectacles in everyday life often find that it is uncomfortable to wear safety glasses at work. They only wear them because it is the rule in the workplace although they recognise that the rule is applied to save them from injury. Over time the action of wearing safety glasses develops into an attitude that wearing them is 'good'. This attitude change can be seen when the same people begin to wear eye protection when they do jobs at home.

The model implies that attitude and behaviour are linked such as to influence and reinforce each other. Psychologists began to look at behaviour as a subject itself rather than as merely an indicator of internal states of mind (i.e. attitudes) following early work by Skinner.[1] This approach has led to behavioural analysis and behavioural modification. By focusing on behaviours, accidents can be prevented. In turn this reinforces an attitude toward a safe work environment and a culture of zero tolerance.

5.2 Principal elements of behaviour modification

Behaviour shaping is a function of management. Employing people to undertake tasks for the benefit and prosperity of the enterprise for which they receive a reward (an income) is itself behaviour shaping.

Behaviour shaping is a function of government. In this case, establishing laws requiring that, for example, citizens wear seat belts when driving with penalties imposed if they are caught not doing so, is behaviour shaping.

Behaviour changing programmes, however, seem most effective when feedback occurs which shows the positive consequences of the safe behaviour. Typical areas of work where behaviour modification to improve safety can be successful include the wearing of personal protective devices, the proper handling of materials, the use of safe working methods around dangerous machines, and housekeeping.

Researchers such as Komaki *et al.*[2,3,4] and Suzler-Azaroff[5] consider that the highlighting of consequences when the desired safety behaviour occurs stimulates the adoption of safe work practices. They also promote the idea that feedback when the desired behaviour occurs is itself a motivational strategy.

5.2.1 The performance management approach

Management gurus have been active in exploring techniques which will improve performance of groups and individuals to achieve business goals. One approach has been termed 'Performance Management'[6], which considers four responses to a behaviour. These responses are termed 'positive', 'negative', 'punishment' and 'extinction'.

In 'Positive Reinforcement' the individual receives something that is wanted or valued after the proper behaviour is completed. Reinforcements of this type encourage the behaviour to be repeated in the future.

'Negative Reinforcement' encourages a desired behaviour when the consequence is *removed*. People will work to avoid certain outcomes such as reprimand, suspension or dismissal. They will choose to repeat a behaviour which avoids or escapes this sort of negative outcome, thereby making the required behaviour more likely to be repeated. 'Punishment' reinforcers aim to decrease the likelihood that the behaviour will be repeated. This type of reinforcement employs the giving of an unacceptable response such as criticism or the allocation of undesired work as a means of reducing the recurrence of the undesirable behaviour.

'Extinction' is a type of consequence in which an outcome desired by an individual is removed following a behaviour and is withheld each time that behaviour occurs with the intention of reducing the occurrence of the behaviour. *Figure 5.2* summarises these behaviour reinforcers.

In everyday life, both in the family and at work, all four types of reinforcement are used. Positive reinforcement is generally viewed as being the most effective motivator to achieve work-related and safety goals since it is often 'free' because a few words of praise or encouragement may be all that is required. Furthermore it often outlasts the presence of the manager or supervisor who gives it, thus making the (safe) behaviour more likely to continue.

This approach has also been applied to safety situations by Krause *et al.*[7].

The key concept is that behaviours are mostly shaped by the expected consequences of that behaviour rather than by anything else. The model put forward is outlined in *Figure 5.3*.

An 'antecedent' is an event which initiates a visible behaviour. A 'consequence' is the outcome of that behaviour. Whilst both antecedents and consequences have an effect on behaviour, the consequences are more powerful in exerting control over and directly influencing behaviour. Antecedents, on the other hand, control behaviour indirectly, largely because they serve to predict the consequences. An example of this theory is to be found in *Figure 5.4*.

BEHAVIOUR

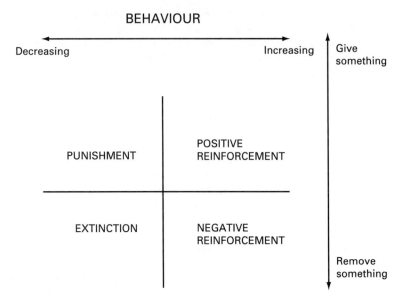

Figure 5.2 Summary of behaviour reinforcers

Further work in this area has shown that there are a number of features which make consequences stronger behaviour modifiers to groups or individuals than others.

The first feature is *Timing*. A consequence that follows on quickly from a behaviour is far more effective than one which occurs after a delay (i.e. later).

The second feature is *Reliability*. A consequence that with certainty will follow a behaviour is more effective than one which may or may not follow that behaviour.

The third feature is the *Nature* of the consequence. When the individual or group feel they gain from the consequence, i.e. it is positive, the effect

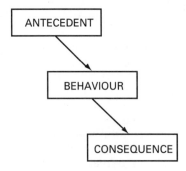

Figure 5.3 The ABC model

Why you do what you do when the telephone rings!

In 'ABC' terms the sequence of events when the telephone rings is as follows:

- Antecedent – the telephone rings
- Behaviour – you answer the telephone
- Consequence – you talk to the caller

However, if the phone rings regularly around the time when the children have returned from school and your experience is that in most cases the calls are for the children, you as the parent soon desist from the Pavlovian response of answering it. Instead you call the children to answer it!

Whilst the antecedent still occurs (the telephone rings) your behaviour to it has been conditioned by the anticipated consequence (the caller will want to talk to the children).

Figure 5.4 An example of antecedents, behaviours and consequences

is more powerful than when they lose, i.e. a negative consequence. There are more problems with this feature than the others.

A positive feature to one person may not be seen as such by another. It can be dependent upon national or local culture. Some people may respond to a simple 'well done' but others may only respond to something far more tangible such as a gift. Problems can also arise if one person's behaviour receives a different level or quality of response from another. For these reasons the positive consequence should be consistent and an appropriate token rather than a chancy lottery win! Notwithstanding this difficulty, a positive response to a behaviour does stimulate a repeat of the behaviour.

When a consequence is imbued with these three characteristics, viz. 'soon', 'certain', and 'positive', it is an effective motivator to achieve the required behaviour. In contrast a consequence which is 'late', 'uncertain', and 'negative' is a weak motivator for achieving the desired behaviour but it is not totally insignificant. At least the behaviour has been recognised and not ignored!

Equally, applying two or even just one feature to the consequence has intermediate levels of influence on the behaviour.

The whole point of a behavioural analysis is to identify consequences which will reinforce the behaviour that is wanted.

The starting point for applying this theory in the workplace is to identify a specific behaviour and analyse it. *Table 5.1* looks at behaviour common to many workplaces, namely the failure by the operator to adjust a computer workstation for individual use. The analysis proceeds by first listing the possible reasons why the current behaviour should occur as it does. These are the antecedents. The consequences of the behaviour are also listed and analysed to establish the features they possess.

It can be seen that the avoidance of musculoskeletal injury or eyestrain are weak consequences because they are 'late', 'uncertain' and 'negative' in nature. The other consequences arising from not adjusting the

Table 5.1 ABC analysis of computer stations

Antecedent	Behaviour	Consequence	s/l	c/u	+/–
Eagerness to use the workstation	Failure to adjust the computer station before use	Saves time	s	c	+
		Musculoskeletal injury	l	u	–
Inadequate training on the need to adjust the workstation		Eye strain	l	u	–
Lack of awareness of the chronic injury potential		No immediate ill effects	s	c	+
Anticipation of zero consequence		Not seen by colleagues to be 'fussy'	s	c	+

s/l – soon/late   c/u – certain/uncertain   +/– – positive/negative

workstation are 'immediate', certain', and 'positive' to the individuals and thus reinforce the unwanted behaviour. Whilst it is possible to debate each component the overall balance of consequences is disproportional and in favour of reinforcing the unwanted behaviour, namely the failure to adjust the chair height, or alter the screen tilt, or draw the blinds so as to prevent glare from windows etc.

The next step in the process is to state in precise terms the observable behaviours which are desired. Then the antecedents and consequences which influence the desired behaviour can be added. This is demonstrated in *Table 5.2*.

The antecedents are likely to focus around education and training followed by ongoing reminders. The consequences would include comments and intervention by supervisors and fellow workers. For the consequences to be effective, supervisors would need regularly and frequently to observe and comment on the individual's use of, and performance at, the workstation. The comments should be positive and approving when the desired behaviour has occurred. Commenting only when the desired behaviour has not occurred is far less effective.

Reducing accidents in the workplace requires that the performance management approach is applied to all unsafe behaviours. Clearly this is a mammoth task that should be approached systematically. It can be achieved either by an analysis of the accident data (i.e. using historical data) or by using Job Hazard Analysis of tasks undertaken (i.e. using predictive data). In either case the analysis is looked at from a behavioural perspective.

Table 5.2 Revised ABC analysis of computer workstations

Antecedent	Behaviour	Consequence	s/l	c/u	+/–
Understanding of injury potential		Observation and comment by supervisor	s	c	+
Training in use	Computer workstation adjusted by user before use				
Expectation of comment by supervisor		Expectation of colleagues	s	c	+
Expectation of ongoing reminders to use workstation correctly		Observation and comment by colleagues	s	c	+

s/l – soon/late   c/u – certain/uncertain   +/– – positive/negative

Using the accidents analysis data, the first step is to group it by task and by area, e.g. 'fork-lift truck accidents in the distribution area'. The second step is to examine how each accident occurred and to identify the significant behaviours which contributed to the accident. This list is variously termed the 'critical behaviour list' or the 'key behaviour list' for fork-lift truck accidents in the area. The third step is to analyse each of the critical behaviours identifying their antecedents and consequences, and the features of each consequence (the 'ABC' analysis). The fourth step is to state the desired behaviour which would avoid the accident and to give it consequences which will reinforce the use of the desired consequences. At this stage this analysis is complete.

Job hazard analysis begins by examining the task and listing the desired behaviours to accomplish it safely. Each desired behaviour is examined and the necessary antecedents and consequences added.

The analysis is the first part of the programme which then has to be implemented. In practice this means that a number of observers have to be trained. Their task is to understand the safety critical behaviours for the workplace and to become skilled in identifying them.

The observers audit the work area recording the number of safety critical behaviours observed and the number of unsafe behaviours observed to produce a 'percentage safe behaviour' score as follows:

$$\% \text{ Safe behaviour score} = \frac{\text{number of safe behaviours observed}}{\text{total number of behaviours observed}}$$

This data is plotted and posted in the workplace so that the workgroup is encouraged to work toward a rising trend. A typical graph of the results is shown in *Figure 5.5*.

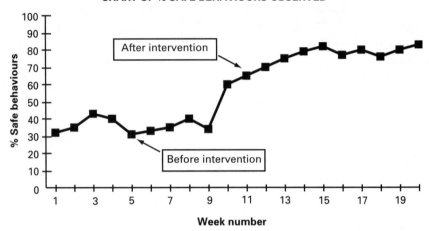

CHART OF % SAFE BEHAVIOURS OBSERVED

Figure 5.5 Typical effect of behaviour intervention process in the workplace

The strength of the process is more than an analysis of actions in the workplace and observations of activity. It works best where the employees take a leading role in managing and implementing it. Employees thus undertake the analysis of behaviours, add the consequences necessary to achieve the safe behaviours and subsequently audit each other. By this means greater commitment to improving safety occurs. In addition the workgroup often know how a job is actually done (as opposed to what the procedure for it says) and is in the best position to draw up the list of desired safety behaviours in the first place and to monitor compliance with it.

When accidents occur they are analysed to see what might have gone wrong. It may be that the safety critical behaviours were not identified correctly in the first place. Alternatively the analysis might reveal that the consequence modifiers are ineffective and have to be rethought.

5.2.2 The structural feedback approach

Performance management requires that consequences are first analysed and then restructured to encourage the preferred (safe) behaviours. The chart demonstrating the improvement in the percentage of safe behaviours is a feedback tool demonstrating the gains made.

Other work by Cooper *et al.*[8,9,10,11] places greater emphasis upon the feedback process. In particular they stress that publicly displaying a chart showing how well, or otherwise, a group of employees is doing in relation to the areas of safety in which improvement is sought is itself a very powerful agent for change. Consequently it becomes important that feedback charts are posted prominently and are regularly updated. In order to achieve this it is necessary that managers adopt a particular role, namely:

1 Champion the behavioural process and inform his workpeople of it and of his support for it.
2 Encourage employees to become active in the process especially as observers.
3 Allow employees the time to be involved in the training and meetings needed for goal setting.
4 Allow each observer one observation session each working day. An observation session should last no longer than 20 to 30 minutes.
5 Be committed to attend goal setting sessions with the observers thereby demonstrating his support.
6 Praise employees who work safely.
7 Encourage employees to reach the safety goals.
8 Arrange for senior managers to visit the workplace each week to encourage the safety improvement effort.

The observers, who are members of the workgroup, commence their training by analysing local accident data. They identify contributory factors for each accident and subdivide them into observable behaviours or situations which are safe or unsafe. These observable data form the basis of a checklist. Emphasis is placed upon gaining agreement from the workforce that the items that form the checklist of behaviours are valid. This is an important step as the workforce is assessed and scored against the list that has been generated. The process of gaining agreement is itself a type of feedback which seeks to gain involvement of and ownership by employees of the safety programme.

Scoring takes the conventional form of making observations in the workplace of safe and unsafe behaviours to generate a 'percentage safe behaviour score'. The data are charted and posted visibly in the workplace. Feedback of the data is not the only emphasis. The employees are asked by their observers to establish their own goals and subgoals against which the performance is measured

5.2.3 Behaviour observation and counselling techniques

Any behaviour modification technique must involve an interaction with people. As an accident can occur at almost any time and the consequences in terms of injury outcome are not predictable, concentrating on a list of identified safety critical behaviours can have the following limitations:

1 The critical behaviour list may be incomplete.
2 Behaviours may appear on the list as a result of a perception of, rather than an analysis of, an actual risk. This is more likely if the list has been compiled from a job hazard analysis.
3 As the size of the list grows to encompass more behaviours (a result of ongoing accident experience and the desire to eradicate all accidents by adding more safety critical behaviours) the whole system can become unwieldy because too may behaviours are included in the observation process.

4 The very existence of a list may limit the focus of employees and observers to only those behaviours which are on the list. This becomes a more significant problem if observers are under pressure to complete a quota of observations per week or per month. Under these circumstances the objective can alter subtly from one of using the technique to reduce accidents to becoming merely an exercise in completing a checklist. The resultant quality fall-off which takes place can undermine and discredit the entire effort.

Other approaches have been developed such as the DuPont Safety Training Observation Program ('STOP') or their similar 'Safety Management Audit Programme'[12]. In both these programmes the approach tends to be less analytical in defining prescribed unsafe behaviours with a different emphasis that requires a management top-down approach in which one level of manager, having been taught the process, subsequently teaches the next subordinate level. The emphasis is upon observation of employee behaviour and immediate counselling of the observed employees. Implementation of the process is through members of line management from team leaders to senior managers. Each undertakes a workplace safety behaviour audit to an agreed schedule. For example, a team leader of a large workgroup may be expected to undertake a daily audit, middle managers may do an audit each week, and senior managers and directors an audit each month. The training they receive assumes a degree of knowledge of the workplace and the hazards it contains. This is not unreasonable given that many managers will have several years' experience and knowledge of the work areas. Furthermore they are not necessarily expected to know in detail how safe working on each job should be achieved. They are expected, however, to recognise how injury might occur.

The training emphasises the skill in observing people as they work and learning to approach and discuss safety with them in a constructive manner. This applies to employees who are observed working safely as well as those working unsafely. In the former case discussion can commend the safe behaviour and be widened to encompass other tasks the employee might do, seeking out any safety concerns arising from them. The very fact that a person in authority is discussing safety issues with the employee is of great importance in raising awareness and commitment to accident-free working. Thus in these programmes, there is greater emphasis on observation and immediate intervention than on observation, completion of a checklist, and the posting of a chart in the workplace. Nevertheless as employees do voice their concerns they have expectations that remedial measures will be taken. Feedback in this case often takes the form of a list of actions identified from the audits and a rolling calculation of the percentage completed.

5.3 The future for behavioural processes

The experience of many established companies is reflected in Figure 5.6.

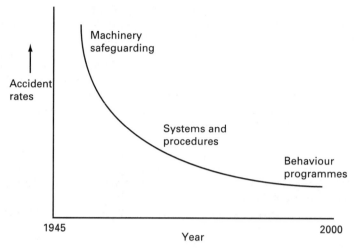

Figure 5.6 Safety emphasis and post-war accidents trends

Accidents have been decreasing over the last fifty years but are tending to reach a plateau. In early post-war years the emphasis remained on physical safeguarding. This gave way to the systems-of-work approaches in the 1970s which continues today in the form of risk assessments. The behaviour-based approach is becoming increasingly important as a technique to improve safety at work for two reasons. First, the other approaches have been implemented but there still remains room for considerable improvement. Second, the nature of work is changing. The emergence of service sector employment and the increasing freedom from a conventional workplace, which modern communication and computer systems allow, have caused written procedures and close supervision to become less effective means of exerting control. Effective safety measures will emerge from better education and training and the success of motivational theories to change attitudes and behaviour.

Whilst an understanding of the links between attitude, behaviour and the consequences of the behaviour are still developing, enough is known to establish that certain key elements of a behaviour approach are more likely to yield success. These key elements include:

1 Management commitment. This should take the form of providing encouragement, being supportive and offering coaching to the work-group. A 'command and control' approach is not a recipe for lasting success.
2 Workforce involvement. It is essential that the employees, who themselves suffer the accidents, are thoroughly immersed in the process. Their involvement is crucial because they are the people who know the unsafe acts that are committed and the reasons for doing them. There may be reasons, which managers and supervisors only dimly understand if at all, why employees take unsafe short cuts to

finish a job. Clearly the workgroup themselves are in the best position to know the antecedents and consequences which operate and how they can be adjusted to promote safe working.

3 Effective feedback is essential and should be twofold in content. First, there should be immediate interaction with the employee observed irrespective of whether safe or unsafe working is noticed. Second, there should be feedback to the workgroup as a whole based upon the agreed checklist of safety critical behaviours. This behaviour score should be compared to goals agreed by the workgroup and progress posted on a chart in the workplace.

A manifestation of an effective behavioural programme is likely to be visible in the form of an exemplary standard of housekeeping. Poor housekeeping must be viewed as the 'detritus of human behaviour'. However, high standards of housekeeping will only be achieved when:

- every item in the workplace has an assigned place
- everyone who works in the workplace knows where everything should be
- everyone who works in the workplace is motivated to place every item where it should be even if they themselves didn't leave it in the wrong place.

Finally just one member of the workgroup can leave just one item in the wrong place just once and in so doing let the whole team down. When humble housekeeping is seen from this perspective it assumes a different dimension and becomes a good indicator of safety performance.

Behavioural processes may have limitations as to their effectiveness. Most examples come from areas where the hazards are acute and understood and the behaviour is observable, such as the wearing of personal protective equipment. It may be harder to apply these techniques to work systems such as 'permits-to-work' or where the risk is low in probability but high in consequence and where precautions are taken but effects not immediately visible. An example might include the maintenance of pressure systems where poor work may not become visible until an explosion occurs much later.

The objective of a behaviour-based approach is to change work habits for the better. There is evidence to show that success does result from such programmes. However, they require considerable resources and commitment which might not always be forthcoming. Little work has been undertaken to show if improvements in performance continue to accrue or can even be maintained when resources are scaled down, for example, by undertaking less frequent audits. It may be that the effort can be scaled down because attitudes have effectively and permanently altered in favour of ongoing safe working. The evidence from major companies which have reputations for continuously superior safety performance seems to be that the *formality* can be reduced but an ongoing management focus must continue. In practical terms this involves being seen to be committed to safe working, walking the workplace, coaching employees into safety behaviours, and encouraging their full participation into what is a stated major workplace value, namely an ongoing reduction in accidents and ill-health at work.

References

1. Skinner, B.F., *The Behaviour of Organisms*, Appleton-Century-Crofts, New York (1938)
2. Komaki, J., Barwick, K.D. and Scott, L.R., A behavioural approach to occupational safety: Pinpointing and reinforcing safe performace in a food manufacturing plant. *J. Appl. Psychol.*, **63**: 434–445 (1978)
3. Komaki, J., Heinzman, A.T. and Lawson, L., Effect of training and feedback: component analysis of a behavioural safety program. *J. Appl. Psychol.*, **65**: 261–270 (1980)
4. Komaki, J.D., Collins, R.L. and Penn, P., The role of performance antecedents and consequences in work motivation. *J. Appl. Psychol.*, **67**: 334–340 (1982)
5. Suzler-Azaroff, B., The Modification of Occupational Safety Behaviour. *J. Occupational Accidents*, **9**: 177–197 (1987)
6. Daniels, A.C. and Rosen, T.A., *Performance Management: Improving Quality and Productivity through Positive Reinforcement*. Performance Management Publications, Inc., Tucker, Georgia (1987)
7. Krause, T.R., Hidley, J.H. and Hodsen, S.J., *The Behaviour-Bases Safety Process*. Van Nostrand Reinhold, New York (1990)
8. Cooper, M.D., Makin, P.J., Phillips, R.A. and Sutherland, V.J., Improving safety in a large, continuous shift, production plant using goal setting and feedback: benefits and pitfalls. Brit. Psychol. Soc. Annual Occ. Psychol. Conference, Brighton, Jan. 3–5 (1993)
9. Cooper, M.D., Goalsetting for safety. *The Safety and Health Practitioner*, November 1993, 32–37.
10. Cooper, M.D., Implementing the behaviour based approach, a practical guide. *The Safety and Health Practitioner*, November 1994, 18–23.
11. Cooper, M.D., Phillips, R.A., Sutherland, V.J. and Makin, P.J., Reducing accidents using goal setting and feedback: A field study. *J. Occ. & Org. Psychol.*, **67**, 219–240, (1994)
12. Safety Training Observation Program. E.I. du Pont de Nemours and Company, Wilmington, Delaware, 1989.

6

Practical safety management

J. E. Channing

6.1 Introduction

Health, safety and environmental issues are often seen as a vast mixture of complex and simple matters hidden in a confusing fog of sectional interests, law and unsympathetic regulators! It is often forgotten that the objective is simply to prevent people from being injured or suffering ill-health from the activities of the enterprise. A jargon has emerged full of 'risk assessments', 'safe systems of work' and 'reasonably practicable options'. The straightforward approach – 'how can we be hurt and what can we do about it?' – has been put aside.

Yet as people live longer and their expectations of good health increase it is inevitable that complexity from ever more subtle risks to our well-being increases.

The conundrum facing many managers is to find sensible ways of dealing with these issues. The solutions will vary with the size and nature of the enterprise. A small marketing operation which just uses computers and telephones to operate its business has fewer and different risks from a supermarket business. Other enterprises may be larger and encompass many different types of operations so that each manager must consider hazards which are general to the whole business as well as ones which are special to the part of the business under local control. This situation can arise on larger mixed occupancy sites where, for example, chemical storage facilities are adjacent to the sales operation. This chapter outlines the responsibilities borne by managers and provides insights into techniques and processes which can be used to manage the health, safety and environmental risks effectively and sensibly.

6.2 Legal obligations

The responsibility for managing the safety of employees lies with the owners of the enterprise and their appointed agents, usually the managers of the workplace. This obligation is a constant feature of legislation throughout the world. In the UK it is encoded in the Health and Safety at Work etc. Act 1974 (HSW) and reinforced in the

Management of Health and Safety at Work Regulations 1992 (MHSW). This basic obligation is also explicit or implicit in all other regulations and is a reflection of each citizen's common law duties.

6.2.1 Common law

Lord Maugham articulated the duties under common law in the court case between *Wilsons and Clyde Coal Co. v. English*[1] when he said:

> 'In the case of employment's involving risk it was held that there was a duty on the employees to take reasonable care, and to use reasonable skill, first, to provide and maintain proper machinery, plant, appliances, and works; secondly, to select properly skilled persons to manage and superintend the business, and, thirdly to provide a proper system of working.'

The above statement was published in 1937. In the same case Lord Wright quoted a previous judgement by Lord McLaren in *Bett v. Delmeny Oil Co.* in 1905[2], which said:

> 'The obligation is threefold, the provision of a competent staff of men, adequate material, and a proper system and effective supervision.'

The duties of employers, thus well established in common law, became encoded in criminal law in the 1974 Act where these duties are applied to the extent that is 'reasonably practicable'. The meaning of this phrase was summarised in a common law case by Lord Asquith in his judgement in *Edwards v. National Coal Board*[3]. He said:

> ''Reasonably practicable' is a narrower term than 'physically possible', and seems to me to imply that a computation must be made by the owner in which the quantum of risk is placed on one scale and the sacrifice involved in the measures necessary for averting the risk (whether in money, time or trouble) is placed in the other, and that, if it be shown that there is a gross disproportion between them – the risk being insignificant in relation to the sacrifice – the defendants discharge the onus on them. Moreover, this computation falls to be made by the owner at a point of time anterior to the accident.'

The phrase and the interpretation once more summarised the common law duty of care to take 'reasonable care'. How they have been applied to different accident situations is to be found in many legal publications such as *Munkman's Employer's Liability*[4]. Common themes emerge where factors such as the nature of the hazard, the obviousness of the hazard, the potential consequence to the employee as well as the cost of the control measures must be considered.

6.2.3 Statute law

Now that these common law duties are encoded in HSW, failure to comply is a criminal offence. Additional regulations extend the duties by requiring risk assessments to be undertaken as a means of determining the appropriate control measures.

Examples where regulations requiring risk assessments include MHSW, the Manual Handling Operations Regulations 1992 (MHOR) and the Health and Safety (Display Screen Equipment) Regulations 1992 (DSER).

The practical problem thus faced by employers is to identify the risks to employees and to the organisation, identify applicable regulations, and fulfil the duties to the extent that will be deemed reasonable. The first step is to set out the goals and objectives and the means of achieving them. In some organisations this might be termed a 'vision statement' or a 'mission statement', but in legal terms it is called a safety policy.

6.2.4 The safety policy

HSW, s. 2(3), requires that each enterprise employing over five people must have a safety policy.

The safety policy itself is a statement of safety intent, but it needs to be supported by a statement of the roles and responsibilities of those who implement it and the administrative arrangements which support it.

The statement of roles and responsibilities can be by an organisation chart showing the positions held and the roles played in applying the policy. This section is most important because it places clear responsibilities on people to fulfil the tasks they are given. The smallest enterprises should have little difficulty in placing this responsibility since the main duty holder will be the owner or manager. However, with larger enterprises where employees hold positions of authority, their roles and responsibilities in respect of health and safety must be stated. Large enterprises may thus have to specify the role of directors, division managers, department managers, supervisors, team leaders (i.e. the 'management line') as well as technical or specialist staff.

Defining responsibilities in complex organisations can be difficult. The principles that 'responsibility is exercised according to the span of control applied' or to the 'authority delegated' are useful tests. For example, policies may state that the line manager is responsible for the employees' health and safety. However, the technical specialist may be able to alter process formulations or the sequences of the production cycle and these factors may affect safe working. Other administrative staff may change delivery routes in distribution enterprises or the number of telephone lines to be handled by individuals working in call centres, perhaps leading to an increase in 'keystrokes per hour' on associated computer terminals. The roles and responsibilities of non-line management people need to be considered when responsibilities are allocated.

In cases of enforcement following a serious breach of safety legislation, the enterprise is often punished but there is also a growing trend to single out individuals who have not exercised their responsibilities competently.

The adminstrative arrangements should include reference to particular safety arrangements and rules within the company with reference to applicable standards, an HSE guidance note[5] and the codes of practice produced by industry groups. The administrative arrangements should also refer to the safety of others such as contractors or members of the public.

6.3 Safety management

6.3.1 Management models

The law is a crude management model for safety. It defines a goal (a hazard which it seeks to control) and sets forth a set of criteria to achieve that control. It also motivates employers to apply those controls by threatening punishment for failures to do so. The safety policy can also be seen as another attempt to apply a management model to health and safety.

More recently models for health and safety have emerged which are far more user friendly to the hard pressed manager. The HSE's publication 'Successful Health and Safety Management'[5] provides one of many models to manage health and safety issues within an enterprise. It outlines a system based upon establishing a policy with targets and goals, organising to implement it, setting forth practical plans to achieve the targets, measuring performance against the targets and reviewing performance. The whole process is overlaid by auditing. A schematic representation of this process is given in *Figure 6.1.*

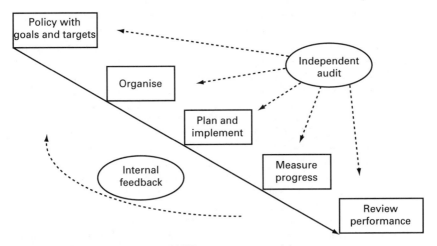

Figure 6.1 Schematic diagram of HSE management model

Another similar model has been developed in BS 8800:1996[6] which chooses as the starting point an Initial Status Review by an independent auditor and the steps that follow are:

1 The occupational health and safety policy.
2 Planning.
3 Implementation and operation.
4 Checking and corrective action.
5 Management review.

The above steps are considered within an overall framework of 'continuous improvement'.

The BSI model has been structured to fit with the international quality management system standard BS EN ISO 9000[7] and the equivalent environmental management system standard BS EN ISO 14001[8]. Applying the BSI model therefore offers the opportunity for companies who have opted for the quality management system to integrate environment and health and safety into one comprehensive management process.

6.3.2 Safety management systems

Many enterprises base their safety management systems on the quality standard or link pre-existing systems to it. In the chemical industry, for example, the 'Responsible Care'[9] programme existed prior to either the HSE guidance or the BSI standard. It has now been crosslinked by the Chemical Industries Association to BS EN ISO 9001.

Regulatory bodies increasingly seek a robust formal management system and managers can expect to be asked, during a site visit by inspectors, to explain how they manage health and safety. It is therefore important that systems are developed which integrate with the responsibilities and the business needs of the enterprise.

A starting point to establish the management system in a company is to consider the financial system of controls which establishes budgetary targets based upon its previous performance and its future aspirations. The overall targets are then broken down to area budgets. Each level of management is given defined targets to achieve and is measured against them. Progress is reviewed regularly at each management level and the whole financial system is independently audited.

The parallel safety system needs to define the safety risks and to establish the effectiveness of the control measures taken. A consideration of accident data and an audit of performance can provide this information and enable an improvement plan for the year to be set.

The management systems described above are high level models and will need to be supported by more detailed programmes. However, they offer little guidance on the process of implementation beyond setting standards and requiring audits and assessments.

In this chapter, an assessment refers to the identification of risks in a local area whereas an audit covers health and safety across the whole organisation with a view to determining its overall health and safety condition.

6.4 Implementing a regulation within a safety management system

An essential component of any safety management system is a mechanism for recognising and implementing new regulations into the working environment. Once identified, the regulation should be broken down into simple discrete components. Choosing the MHOR and the associated Guidance[10] as the example, the process can be applied as follows.

6.4.1 Defining the key steps

In the Regulations, the sequencing of the subsections may be logical from a legal viewpoint, but from a practical manager's point of view they need to be rearranged. They require the employer to assess the risks from manual handling and to develop methods of work that reduce these risks to a minimum. Employees are required to follow the work method.

Step 1   There is no requirement in the Regulations for a list of tasks to be generated. However, planning a compliance programme is difficult without knowing the scope of the task. The first step is to list jobs with a manual handling component together with the specific tasks within those jobs. Groups of similar tasks can be put together for generic assessments. For example, the task of loading supermarket shelves with produce would be one assessment – separate assessments would not be required for stacking baked beans one moment and sugar the next! However, a separate assessment may be required for loading Christmas turkeys into a chest freezer due to the greater weight of the product and the bending component of the task. In effect an informal assessment is being undertaken whilst compiling generic assessments so that like-for-like risks are grouped together.

Step 2   is to undertake the formal risk assessment, as required by reg. 4(1)(b)(i) possibly using the proforma outlined in Appendix 2 of the HSE guidance.

Step 3   is to identify the agreed control measures (or 'remedial actions' in the HSE proforma). It may have emerged from the study that the manual handling task can be eliminated. If so the employer has been able to comply with reg. 4(1)(a). This conclusion, however, may not have been reached until the assessment study has been completed.

Step 4   implements the control measures which have been identified by the assessments. Employers are required by reg. 4(1)(b)(iii) to give employees information on the weight of each load. This can be achieved in different ways. For loads of constant weight it may only be necessary to put the information in a local procedure. However, enterprises which supply a range of goods to a variety of customers may need to state the weight on each package. For off-centre loads, the packaging should indicate the heavier side.

Step 5   is a procedure for reviewing the assessments whenever there is reason to suspect the existing one is invalid, or there has been a significant change in the operation. This step complies with reg. 4(2).

Step 6   implements reg. 5 which requires employees to adopt the safe system of work established by the employer as a result of the assessment. This can be achieved through training to make employees aware of the risks, the proper working methods for the safe performance of the task.

Decoding the regulations into practical and easily understood steps is just the commencement of the implementation process.

6.4.2 Organising the implementation process

In establishing the above steps line managers may be best placed to generate the initial list of jobs and tasks. They will need an understanding of ergonomics and of the regulations. The risk assessments could be undertaken by line managers or team leaders since they usually have a better appreciation of the tasks undertaken in both normal and abnormal circumstances. It becomes their responsibility to find less risky alternatives and hence gain ownership of the whole process. Furthermore, they fulfil a basic tenet of safety law in that the employer, through his line managers, has responsibility for the safety of his employees. Some companies use safety advisers, specialist trainers or medical specialists to undertake the assessments. There are, however, considerable benefits in keeping as much of the assessment and implementation activity 'in the line' rather than offloading them onto a specialist.

The safety adviser's role is to assist in the construction of the implementation process and training those in the team who will undertake the various assessment tasks.

6.4.3 Measuring the progress

'What gets measured, gets done' was a phrase coined by Peters[11]. The phrase sums up the belief that managers respond best to measurement systems, often numerical, to chart progress and thereby ensure plans reach completion. A measurement matrix can be applied to implementing regulations once the steps have been established. An example of a measurement matrix for implementing the manual handling operations risk assessment is presented in *Figure 6.2*.

The matrix has several important features. Some steps are measured so that a 'No' answer attracts a zero score whilst a 'Yes' answer gains the maximum score. Other steps gain a graduated score according to the percentage of the task completed. Another important feature is that not all steps are equally weighted. In general the higher weightings are given to the more important steps or those requiring the greater workload. In the example, step 4 has the heaviest weighting as this is viewed as the single most important step to achieve a safe manual handling workplace.

MANUAL HANDLING REGULATIONS COMPLIANCE CHART

STEP	LEVEL ACHIEVED											SCORE	WEIGHT	OVERALL SCORE
score values	0	1	2	3	4	5	6	7	8	9	10			
1. List of jobs and tasks generated	no										yes		5	
2. % risk assessments completed	0	10	20	30	40	50	60	70	80	90	100		20	
3. List of control measures agreed	no										yes		15	
4. % of control measures implemented	0	10	20	30	40	50	60	70	80	90	100		25	
5. procedure to mark loadweight	no										yes		5	
6. Review process in place	no										yes		10	
7. % employees trained	0	10	20	30	40	50	60	70	80	90	100		15	
8. Training packages available	no										yes		5	
													TOTAL SCORE=	

Figure 6.2 Measurement matrix for the implementation of MHOR

Adjusting the weighting in this manner encourages a focus on the essential elements of complying with the Regulations.

As the implementation plan proceeds, a score for each step is calculated by multiplying the level achieved by its weighting. A copy of the chart can be provided to senior managers each month from which the pace of progress can easily be seen. This can assist in decisions regarding the allocation of resources to maintain progress.

In a multi-departmental enterprise, each area should have its own matrix with summary scores published to spur the laggards to catch up. It acts as a simple but effective motivational and behaviour-shaping tool.

6.5 Safety management and housekeeping

Managers often target 'housekeeping' as an area which needs improvement. Many accidents occur because of poor housekeeping in the form of uncleared spillages, overstacked shelves etc., which indicate a lack of control in the workplace. The achievement of high housekeeping standards is also a key indicator of good safety performance in the area because it is a manifestation of the attitudes that prevail amongst those working there. Good housekeepng requires a well-structured work process, discipline in execution, and motivated employees. For this reason an HSE inspector may use the workplace condition to indicate possible weaknesses in the safety management systems.

Scoring techniques can be used to improve and maintain housekeeping in the work area and an example of a checklist is provided in *Table 6.1*. The checklist identifies the expectations clearly and precisely and should be phrased in such a way that full compliance gains the maximum points. Questions can range from the physical conditions in the workplace to the practical knowledge of the employees. Approaching housekeeping in this way has several benefits. Firstly, the area's requirements are stated clearly. Secondly, because the questions are clear and concise, each employee in the work area can complete the checklist on a rota basis. Thirdly, the checklist can be changed periodically.

Changing the checklist contents needs to be considered carefully so that the desired behaviours and standards are developed and improved. Leaving an item off a well-established checklist may lead to it not receiving attention in the future. On the other hand, simply adding more and more items may make the whole checklist unwieldy. If this occurs, a split checklist may be necessary with each part completed in alternate weeks.

A final point to consider relates to the scores themselves and whether to use 'points' or 'percentages'. In general 'points' are preferred because new items to the checklist add more to the 'maximum points horizon' without diminishing the value of the points scores for the existing items. 'Percentages', where 100% is the goal, do not achieve this, thereby diminishing the relative worth of an existing item on the checklist.

Table 6.1 An example of a scoring checklist for housekeeping

Housekeeping Checklist – *Computer Suite*

Item	Maximum score	Actual score
1 Emergency fire escape routes are unobstructed (*2 routes, 10 points per clear route*)	20	
2 First aid boxes contain correct contents (*2 boxes, 4 points per box*)	8	
3 Work areas clear of trailing leads in walkways (*deduct 2 points per trailing lead*)	10	
4 Electrical leads in area visually inspected and show no sign of mechanical damage (*deduct 4 points per damaged lead*)	12	
5 Computer workstation meets requirements of Regulations (see supplementary list) (*2 workstations, deduct 4 points per defect*)	20	
6 Workstation users have made correct adjustments for their own use (*2 workstations, deduct 4 points per defect*)	20	
Score maximum =	90	actual =

6.6 Assessment techniques

The models briefly reviewed at the beginning of this chapter set an overall framework for managing health and safety issues. Once the system is established there is a requirement for it to be assessed. The role of an assessor can vary. One aspect is to review the management system and procedures to ensure that they can operate as a cohesive structure. It does not need a professional understanding of health and safety to perform this task. Another approach that might be taken is to use a qualified health and safety adviser who checks the system and is able to judge whether safety in the workplace is being achieved. Whichever approach is used the assessment report should list positive features of the system as well as listing 'non-compliances', 'negative findings' or 'corrective action requests' according to the terminology used. The benefit of this approach is that it requires an assessment by a third party.

But there are weaknesses. Even an enterprise which, by any objective standards, is good in terms of health and safety performance may still be given a list of non-conformances. Also the items listed in successive assessments become of increasingly less importance, even trivial, and may reflect the assessor's opinion rather than a requirement of a regulation or internal standard.

A typical response to a third party health and safety assessment is to see it as a target to be aimed for and to do just sufficient to satisfy the assessment. This response can prevail where managers fail to embrace health and safety as central to their function.

Assessments may be undertaken once or twice a year and may follow a 'sampling' technique whereby a few specific areas are examined in depth and conclusions are drawn which can be extrapolated to cover the whole organisation. An assessment may seem like an examination where the result is always 'can do better'! Whilst assessments are necessary and can be beneficial to an organisation they are often approached with apprehension.

6.6.1 Restructuring health and safety assessments

Techniques have been developed to make health and safety assessments an ongoing and central element of a manager's function. Section 6.4 demonstrated how a regulation can be translated into an assessment with a numerical format so that progress towards implementation can be given a score. The same principle of scoring can be applied to embrace all regulatory requirements once they have been implemented with other health and safety aspects of the workplace. Table 6.2 shows a simple set of questions which can be used to maintain compliance with MHOR Each question attracts a points score as indicated.

Table 6.2 Example of audit questions for maintaining compliance with manual handling operations regulations

Audit question	Points available	Points scored
1 Have tasks in the area been reviewed in the last year to identify those with a manual handling component?	10 if 'yes'	
2 State the % of manual handling tasks with formal written risk assessments	50 points for 100% compliance and pro rata	
3 On what % of tasks has the risk been reduced by the control measures to as low as reasonably practicable?	50 points for 100% compliance and pro rata	
4 What % of employees know where the risk assessments are kept and have been made familiar with them?	50 points for 100% compliance and pro rata	
5 Have steps been taken to inform employees of load weights?	20 if 'yes'	
6 Have assessments been reviewed within the last 3 years?	30 if 'yes'	
7 What % of employees have undergone training in handling methods in the last 5 years?	30 points for 100% compliance and pro rata	
8 Is there a procedure in place to review assessments if significant change occurs?	10 points for 'yes'	

Table 6.3 An example of audit questions for safety committees

Audit question	Points available	Points scored
1 Does the area have a joint safety committee?	10 for 'yes'	
2 Has the committee met at least every 3 months in the last year?	10 per meeting up to 40	
3 Does the area manager attend the meetings?	20 per attendance	
4 Does the safety adviser attend the meetings?	10 per attendance	
5 Do employee representatives attend the meetings?	10 per meeting up to 40	

The scoring concept can be applied to each element of the management systems. *Table 6.3* is an example of a question set to manage the performance of safety committees.

The questions are designed to be straightforward and easily understood by everyone. This makes it possible for any employee or a small group of employees to monitor the performance of the system.

The scoring principles for the question sets should give fewer points for simple administrative tasks and greater points for actions by the people involved.

In the safety committee example, the mere existence of a safety committee attracts few points. Furthermore the attendance of the manager outscores the attendance of the safety adviser and employee representatives because the manager has the greater responsibility and plays the key role in safety issues in the workplace.

This is an example of the technique of 'guiding' which can be used to guide the contents of a set of practical actions aimed at improving health and safety performance in the area.

This approach can be applied across the range of health and safety hazards which exist in the work area but it does raise issues of the scores given to dissimilar tasks. For example, one question on fire safety may concern the maintenance of suitable fire escape routes. The administrator of the assessment process must decide how many points should be given to that question compared to a question on labelling of chemicals. There is no absolute answer. Nor need there be. What is important is that each assessment is repeated on a regular basis. Increasing scores for particular areas indicate improving safety standards.

The safety adviser has a key role in establishing a system for ongoing assessments and for deciding on the questions and their scores. Software packages are available and may be preferable to paper systems in some organisations[12,13,14].

Scoring systems of this type permit managers to undertake a self-audit of the performance of their areas of responsibility. However, rather than carry out a comprehensive audit of all areas of responsibility occasionally, greater benefit may be derived by assessing one aspect only each month.

The following section headings have been used with success in some organisations:

1 Health and safety management and administration.
2 Fire, loss and emergencies.
3 Investigation and monitoring.
4 Chemicals and substances.
5 Environment and waste.
6 People.
7 Systems of work.
8 Machinery, plant and equipment.
9 Product safety.

Each section itself can be split into separate parts. For example, section 6 on 'People' can be split into:

6.1   Standard operating procedures.
6.2   Permits-to-work and lock-off systems.
6.3   Manual handling operations.
6.4   Repetitive work.
6.5   Large vehicles.
6.6   Lone workers.

One section should be completed each month over an agreed period of, say, 9 months. In the 3 months following the final assessment the safety adviser can review and agree the scores, thereby providing an independent validation of the self-assessment. It also allows the manager to review his performance and act to improve his score before submitting the final performance score for the year to his director.

With ongoing assessments an opportunity can be taken to alter or add to the question set for the following assessment. The amendments to the assessment enable the enterprise to incorporate newly implemented regulations into the package. It also allows an examination of its performance and identification of the improvements to be made. In one organisation, it was observed that a safety committee in one area was less than effective because it didn't have a procedure for closing actions. A procedure was implemented and additional questions (see *Table 6.4*) were added to the existing assessment form. As the questions set applied to all

Table 6.4 Additional audit questions on safety committees

Audit question	Points available	Points scored
6 Is a list of actions produced with agreed timelines for completion?	20 for 'yes'	
7 What % of actions were completed in the last year within time?	100 points for 100%	

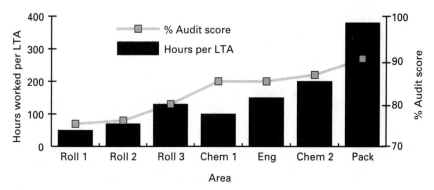

Figure 6.3 Correlation between accident and audit performance

areas of the organisation the deficiency observed in one area was able to be addressed in all areas.

The objective of any assessment process is to reduce accidents at work.

Figure 6.3 shows data from an organisation using the self-assessment system outlined above. Graphs of this sort can demonstrate correlations between lost time accident levels and self-assessment scores. A manager who takes the right actions in a structured approach to health and safety should produce an improved accident performance.

6.7 Proprietary audit systems

Several proprietary audit systems are available. The International Safety Rating System was one of the first comprehensive safety audit systems to be developed. It categorises health and safety issues into 20 management elements and includes a 'Physical Conditions' inspection. From the score obtained an enterprise can be judged to have achieved one of five standard levels or one of five advanced levels. Achievement at the highest level demands a high score in over 600 questions in the management elements and high performance in 'Physical Conditions' (loosely equated to housekeeping). The assessment is undertaken by a third party auditor who has been trained in the system. A successful application of the system also requires that managers are educated to understand its jargon and objectives. Using the system means that comparisons can be made between companies in different business sectors and even in different countries. It has been developed from considerable use over many years to give organisations comprehensive audit processes.

Disadvantages of the system include the scores given to each question and the 'pass marks' required for each level which are set by the system and do not allow flexibility to meet a particular emphasis on which the enterprise wishes to focus. Another problem arises in the application to enterprises that use a different culture and jargon from that employed by the system. Finally, the system does not address specific national legislative requirements since it does not set out to focus on particular national laws but seeks a position above them by asking questions such as 'Does the organisation have copies of relevant legislation and related standards?' It is necessary for the auditor to know the legislation appropriate to the enterprise and exercise personal judgement on the score awarded (up to a maximum of 25 points). Additional supplementary work would have to be undertaken to ensure compliance with specific regulations.

Similar proprietary systems are available from other organisations such as the British Safety Council who offer a 5-star audit system[13]. The scope of the proprietary systems has been developed after considerable experience in many businesses and cultures. They capture all of the key components for achieving superior safety performance. If an enterprise chooses an internally developed self-assessment route, then keeping a proprietary system as a reference and guide for future developments may be useful.

Computer-based systems such as 'CHASE'[14] can provide preloaded audit questionnaires or blank formats into which the organisation's own audit questions can be written. The benefits of writing the audit questions within the organisation is that it will reflect the culture and jargon of the workplace, but the process of setting it up can be time consuming.

6.8 Benchmarking

Benchmarking is an increasingly important tool in safety management. The term 'benchmarking' describes a concept of comparison with and learning from other companies. There is a misconception that benchmarking of safety performance and processes should only take place with companies who are perceived to be superior, i.e. who set the 'benchmark'. However, by undertaking a benchmarking exercise the enterprise is able to obtain a better understanding of its own safety processes. This aspect alone brings benefits before it seeks to identify alternative processes and quality gaps in comparable processes observed in other companies.

A natural starting point for benchmarking exercises is to focus on the elements contained in their audit process. Features included in benchmarking studies include:

- identification of the elements of the safety system
- how the elements fit together
- how management leads and monitors the safety system
- what topics are subject to formal risk assessments, who does them, and what is their quality
- how employees are involved and motivated to play a full part in safety.

From the above it is clear that safety benchmarking is less about a comparison of accident statistics and much more about comparing management processes and their effectiveness. Benchmarking should not be the sole preserve of the safety adviser but should involve line managers who can gain considerable benefit from it. The process compels them to focus their attention on both the subject of safety and their own performance. Safety advisers benefit because it enables them to formulate new improvement strategies which have been proved elsewhere and can be recommended to managers with confidence.

6.9 Involving employees

Regulatory, enforcement and judicial emphasis over many years has focused on the employers' responsibility for safety because the employer is the controlling influence in any enterprise. However, it is increasingly being recognised that there is, and needs to be, greater involvement of employees in improving safety in the workplace. Many companies have streamlined their organisations by removing layers of management and have appointed working 'team leaders' who have greater responsibilities than the 'first line supervisors' of previous years they replace. Team leaders, akin to player managers in sporting terms, are expected to play in the team, coach fellow players, and apply discipline when necessary. They need the ability to be a member of the workgroup, control and direct it, and to be the management authority in the workplace, a taxing and difficult challenge.

Away from traditional working environments, increasing numbers of employees work from home or from their cars taking advantage of modern telecommunication and computer facilities to link with their workplace. In these circumstances it is more than ever important to involve the lone workers in their own health and safety management.

6.9.1 Safety committees

A safety committee is the traditional and effective means of achieving involvement and joint consultation in traditional workplaces. Committees are able to operate most effectively when roles and functions are clearly established. Safety committees which are established at shop floor level need to involve a cross-section of employees and/or their representatives and also representatives from supervision. They are ideally suited to finding local solutions to day-to-day problems. Due to their proximity to the workpeople, committee members should be effective in:

- identifying problems
- developing practical solutions to those problems
- stimulating the commitment of fellow employees to the practical solutions
- promoting awareness and an attitude of mutual assistance.

Solving safety problems is often a matter of applying common sense rather than obeying the law. Once a risk is identified and understood – even chronic health risks or rare catastrophic risks such as explosions – the problem becomes one of shaping behaviour to follow control measures and prevent repetition. Even 'ordinary' accidents such as slips, trips, falls and mishandling, which cause bruises, strains and cuts and form the majority of accidents, would have been avoided if the behaviour of the employee had been different at the time the risk arose. Reaching a consensus on the direction to be taken to improve safety performance should be an objective of any safety committee.

How to influence behaviour at key moments is the subject of much psychology research. Most researchers would agree that awareness of the hazard and an appreciation of the consequences by the individual, if recognised at the moment of risk, would prompt avoiding action on the basis of self-preservation if no other. Members of a shop floor safety committee who are active champions of safety in the workplace – encouraging, motivating and coaching fellow employees – can make a significant contribution to safety standards. Rotating the members of the local safety committee can increase the core of active champions, thus further improving performance. The function of a committee at this organisational level is to be active in the workplace as opposed to discussing safety in meetings.

Safety committees at higher organisational levels also have a role. The terms of reference at this level would be entirely different from shop floor committees and would be concerned with policy and formulating strategies to address safety issues, the implementation of new legislation, monitoring performance, the development and application of safety training programmes and making suitable resources available.

6.9.2 Involving the individual

A challenge to a manager is to involve the individual employees in a structured way to improve their safety contribution. In this he can refer to the general safety expectations outlined in the job description and any specific safety responsibilities associated with the job. Agreed performance monitors should be established wherever possible to measure the contribution made. For example, a sales representative could report each week on the cumulative number of miles driven since the last road accident. Making this data a part of the weekly sales report integrates safety with business. Employees who work with equipment either at home or in the conventional workplace could be asked for their recommendations for improving safety in its use and what changes they, as employees, can make to avoid any of the risks they may have identified. Formalising this question into a regular proforma report from the employee focuses attention and reinforces safety as a joint activity.

6.9.3 Creating the culture

Within every organisation there is an ambience that permeates all levels and is referred to as the 'culture' of the organisation. Its tone is set by the most senior executives and directors and depends on the emphasis they place on particular aspects of the organisation's activities. Commonly it is business orientated; elsewhere it may be financial – the 'bottom line' syndrome. However, a similar phenomenon can occur with safety and it is found that those organisations with a sound safety culture not only have a good safety performance but also high productivity levels.

Lower down in the organisation, ambitious managers seeking high standards of safety need to create a fertile soil in which the flower of safety can grow and blossom. The gardening analogy is appropriate. There are many variables which need to be understood, nutrients need to be applied, weeds removed and the garden needs to be constantly tended.

Achieving outstanding safety performance is no less complex but the blend of components required is probably less well understood. It is employees who mostly suffer the injuries so that understanding their perceptions and attitudes can help focus on concerns which they see as constraints. Structured survey tools to establish these perceptions and attitudes have been developed by psychologists such as Stanton and Glendon[15] and another is available in a publication from the HSE[16].

This latter publication asks 71 questions covering the following 10 factors:

1 Organisational commitment and communication.
2 Line management commitment.
3 Supervisor's role.
4 Personal role.
5 Workmates' influence.
6 Competence.
7 Risk-taking behaviour and some contributory influences.
8 Some obstacles to safe behaviour.
9 Permit-to-work.
10 Reporting of accidents and near misses.

In following the process, managers, supervisors and their workplace employees each provide their opinion on these 10 topics. This permits the views which managers hold to be compared with the views of other levels in the organisation. The survey can be undertaken in other areas of the company and, of course, at other dates enabling comparison across organisations and through time.

Measurement is only the first step in using survey techniques. Collected data, including accident and incident data, should be used for discussion with employees. Utilising the data in this way is as important as getting it in the first place. Involvement of employees in this way raises expectations, employees expect more from managers and managers expect a greater contribution from employees. Satisfying these expectations through constructive engagement requires considerable interpersonal skill.

6.9.4 Avoiding conflict

Conflict in the workplace over health and safety issues usually results from differences in perceptions, whether of the interpretation of rules, of risk or inadequate responses to safety issues.

Studies have shown that disparate groups of people view risks differently, a situation that rises naturally from differences in their backgrounds, education and position in society. The fact that it affects safety is not surprising since safety issues are a social phenomena.

At the beginning of industrialisation in western societies, most concern was focused on machinery accidents and public health issues involving diseases such as cholera. Few were then concerned about the effects of chemicals or smoking because they had not been identified as risks. Today, popular concern rises rapidly with risks that are identified but not understood, such as radiation or BSE and the newly developed substances whose effects could be chronic and cumulative. In the workplace, managers have the primary responsibility for safety so their perception should prevail if differences arise. However, their view must be founded upon good data and sound judgement and then be well communicated if it is to have credibility with employees. Managers should realise that employees have access to up-to-date technical data through their unions.

Conflict can also arise from inadequate or total lack of responses to safety issues that have been raised. Managers who do not respond to employees' safety concerns invite their reaction. Employees can also be unresponsive to safety requirements. They may not follow procedures, or properly wear personal protective equipment, or check emergency stops before using equipment as required. There should be agreement that violation of local rules attracts disciplinary sanctions. A functioning safety committee with actions allocated to appropriate people with agreed timelines for responses can do much to avoid these situations. As can a recording system that shows the number of actions completed out of the total number of safety items raised. Conflict is avoided when all make a positive visible contribution appropriate to their function in the organisation.

6.10 The role of specialists

Qualified health and safety advisers can be used in the same way as accounting specialists are employed in the organisation. The role of the health and safety adviser lies in providing advice, in monitoring performance, audit, data analysis, programme research and development, tracking legislation and emerging public concerns, and the provision of specialist services such as education and training. In this type of environment the health and safety adviser is expected to have knowledge of law, regulations and safety technologies as a minimum requirement. However, considerable additional value is added to the role in the organisation by acting as a coach, trainer and motivator to employees and managers at all organisational levels.

When there is a practical safety task to be done the health and safety adviser should seek first to persuade a person in the line management to do it. For example, a Job Hazard Analysis (JHA) requires detailed knowledge of the production process and it would be much more effective if carried out by a production team leader who may need training in the JHA technique. Not only would the team leader be best placed to understand the normal work process but also be more likely to identify abnormal operations. He would also know how work is actually done compared with what the existing written procedure says. Recommendations for changes would carry more weight with the workforce when produced by someone, known to them, who has current practical knowledge of the work.

6.11 Conclusion

Safety legislation has been on the statute book for many years but the development of safety management as a subject is a recent innovation. It is now recognised that the levels of safety expected by employees and the general public cannot be achieved without utilising safety management strategies. So much has been written about the subject that it is sometimes forgotten that conventional management techniques are just as applicable to safety as to any other aspect of business. In particular the emergence of quality management processes is highly relevant. The exact application of each element of a cohesive practical safety management system depends on the area of business and the aspirations of the organisation itself. Effective safety management does not occur by chance. It arises out of a clear understanding of obligations and the application of considered strategies and techniques. Managing safety does not exist in isolation from other aspects of the business. An enterprise may change direction in terms of its business plan, its products or markets which in turn result in increases or reductions in the number of employees, relocation of premises and adopting changing technologies. In all these circumstances the obligations to the safety of its employees, customers and the public at large remain but it is only the style and techniques applied that differ. Recognising changing operational environments and adjusting the management techniques to suit are essential to improving safety. The principles outlined in this chapter remain constant. The techniques described offer flexibility. Together they offer a pragmatic approach to safe working for the busy manager.

References

1. Wilsons and Clyde Coal Co. Ltd. *v.* English (1938) AC 57 (HL)
2. Bett *v.* Dalmey Oil Co. (1905) 7F (Ct of Sess.) 787
3. Edwards *v.* National Coal Board (1949) 1 KB 704, (1949) 1 All ER 743
4. Munkman, J., *Employer's Liability at Common Law*, 12th edn. Butterworths, London (1996)
5. Health and Safety Executive, Guidance booklet no. HSG 65, *Successful health and safety management*, HSE Books, Sudbury (1997)
6. British Standards Institution, BS 8800:1996, *Occupational health and safety management systems*, BSI, London (1996)
7. British Standards Institution, BS EN ISO 9000, *Quality systems. Specification for the design/ development, production, installation and servicing*, BSI, London (1994)
8. British Standards Institution, BS EN ISO 14001, *Environmental management systems. Specification with guidance for use*, BSI, London (1996)

9. Chemical Industries Association, *Responsible Care*, CIA, London
10. Health and Safety Executive, Legal publication no. L23, *Manual Handling. Manual Handling Operations Regulations 1992, Guidance on the Regulations*, HSE Books, Sudbury (1992)
11. Peters, T.J. and Waterman, R.H., *In Search of Excellence*, Harper & Row, New York (1982)
12. Bird, F.E. and Germain, G.L. *Practical Loss Control Leadership*, International Loss Control Institute, Loganville (1986)
13. British Safety Council, *5 Star Health and Safety Management System Audits*, BSC, London
14. HASTAM, *CHASE*, HASTAM, Birmingham
15. Stanton, N. and Glendon, I., *Safety Culture Questionnaire*, Griffith University, Australia and University of Southampton, England (1996)
16. Health and Safety Executive, *Health and Safety Climate Survey Tool*, HSE Books, Sudbury (1997)

The Institution of Occupational Safety and Health

The Institution of Occupational Safety and Health (IOSH) is the leading professional body in the United Kingdom concerned with matters of workplace safety and health. Its growth in recent years reflects the increasing importance attached by employers to safety and health for all at work and for those affected by work activities. The Institution provides a focal point for practitioners in the setting of professional standards, their career development and for the exchange of technical experiences, opinions and views.

Increasingly employers are demanding a high level of professional competence in their safety and health advisers, calling for them to hold recognised qualifications and have a wide range of technical expertise. These are evidenced by Corporate Membership of the Institution for which proof of a satisfactory level of academic knowledge of the subject reinforced by a number of years of practical experience in the field is required.

Recognised academic qualifications are an accredited degree in occupational safety and health or the Diploma Part 2 in Occupational Safety and Health issued by the National Examination Board in Occupational Safety and Health (NEBOSH). For those assisting highly qualified OSH professionals, or dealing with routine matters in low risk sectors, a Technician Safety Practitioner (SP) qualification may be appropriate. For this, the NEBOSH Diploma Part 1 would be an appropriate qualification.

Further details of membership may be obtained from the Institution.

Appendix 2

Reading for Part I of the NEBOSH Diploma examination

The following is suggested as reading matter relevant to Part 1 of the NEBOSH Diploma examination. It should be complemented by other study.

Module 1A:	The management of risk	Chapters	**2.1–all** **2.2–paras. 8–11** **2.3–all** **2.4–paras. 1–3** 3.8–paras. 1–6 4.7–para. 11
Module 1B:	Legal and organisational factors	Chapters	1.1–all **1.2–all** 1.3–paras. 1–6 1.7–para. 2 1.8–all **2.2–paras. 13 and 14** **2.6–paras. 1–4**
Module 1C:	The workplace	Chapters	1.7–para. 2 3.6–all 3.7–all 4.2–all 4.4–paras. 1–8 4.6–paras. 2 and 4 4.7–paras. 1, 2, 7 and 11
Module 1D:	Work equipment	Chapters	4.3–all 4.4–all 4.5–all
Module 1E:	Agents	Chapters	3.1–all 3.2–all 3.3–all 3.5–paras. 1–6 3.6–all 3.8–paras. 4–7 4.7–paras. 1–4
Module 1CS:	Common skills	Chapter	**2.5–para. 7**

Additional information in summary form is available in *Health and Safety . . . in brief* by John Ridley published by Butterworth-Heinemann, Oxford (1998).

List of abbreviations

ABI	Association of British Insurers
AC	Appeal Court
ac	Alternating current
ACAS	Advisory, Conciliation and Arbitration Service
ACGIH	American Conference of Governmental Industrial Hygienists
ACoP	Approved Code of Practice
ACTS	Advisory Committee on Toxic Substances
ADS	Approved dosimetry service
AFFF	Aqueous film forming foam
AIDS	Acquired immune deficiency syndrome
ALA	Amino laevulinic acid
All ER	All England Law Reports
APAU	Accident Prevention Advisory Unit
APC	Air pollution control
BATNEEC	Best available technique not entailing excessive costs
BLEVE	Boiling liquid expanding vapour explosion
BOD	Biological oxygen demand
BPEO	Best practicable environmental option
Bq	Becquerel
BS	British standard
BSE	Bovine spongiform encephalopathy
BSI	British Standards Institution
CBI	Confederation of British Industries
cd	Candela
CD	Consultative document
CDG	The Carriage of Dangerous Goods by Road Regulations 1996
CDG-CPL	The Carriage of Dangerous Goods by Road (Classification, Packaging and Labelling) and Use of Transportable Pressure Receptacle Regulations 1996
CDM	The Construction (Design and Management) Regulations 1994
CEC	Commission of the European Communities

CEN	European Committee for Standardization of mechanical items
CENELEC	European Committee for Standardisation of electrical items
CET	Corrected effective temperature
CFC	Chlorofluorocarbons
CHASE	Complete Health and Safety Evaluation
CHAZOP	Computerised hazard and operability study
CHIP 2	The Chemical (Hazard Information and Packaging for Supply) Regulations 1994
Ci	Curie
CIA	Chemical Industries Association
CIMAH	The Control of Industrial Major Accident Hazards Regulations 1984
CJD	Creutzfeldt–Jacob disease
COD	Chemical oxygen demand
COMAH	The Control of Major Accident Hazards Regulations (proposed)
COREPER	Committee of Permanent Representatives (to the EU)
COSHH	The Control of Substances Hazardous to Health Regulations 1994
CPA	Consumer Protection Act 1987
CTD	Cumulative trauma disorder
CTE	Centre tapped to earth (of 110 V electrical supply)
CWC	Chemical Weapons Convention
dB	Decibel
dBA	'A' weighted decibel
dc	Direct current
DETR	Department of the Environment, Transport and the Regions
DG	Directorate General
DNA	Deoxyribonucleic acid
DO	Dangerous occurrence
DSE(R)	The Health and Safety (Display Screen Equipment) Regulations 1992
DSS	Department of Social Services
DTI	Department of Trade and Industry
EA	Environmental Agency
EAT	Employment Appeals Tribunal
ECJ	European Courts of Justice
EC	European Community
EEA	European Economic Association
EEC	European Economic Community
EcoSoC	Economic and Social Committee
EHRR	European Human Rights Report

EINECS	European inventory of existing commercial chemical substances
ELF	Extremely low frequency
ELINCS	European list of notified chemical substances
EMAS	Employment Medical Advisory Service
EN	European normalised standard
EP	European Parliament
EPA	Environmental Protection Act 1990
ERA	Employment Rights Act 1996
ESR	Essential safety requirement
EU	European Union
eV	Electronvolt
EWA	The Electricity at Work Regulations 1989
FA	Factories Act 1961
FAFR	Fatal accident frequency rate
FMEA	Failure modes and effects analysis
FPA	Fire Precautions Act 1971
FSLCM	Functional safety life cycle management
FTA	Fault tree analysis
GEMS	Generic error modelling system
Gy	Gray
HAVS	Hand-arm vibration syndrome
HAZAN	Hazard analysis study
HAZCHEM	Hazardous chemical warning signs
HAZOP	Hazard and operability study
hfl	Highly flammable liquid
HIV+ve	Human immune deficiency virus positive
HL	House of Lords
HMIP	Her Majesty's Inspectorate of Pollution
HSC	The Health and Safety Commission
HSE	The Health and Safety Executive
HSI	Heat stress index
HSW	The Health and Safety at Work, etc. Act 1974
Hz	Hertz
IAC	Industry Advisory Committee
IBC	Intermediate bulk container
ICRP	International Commission on Radiological Protection
IEC	International Electrotechnical Committee (International electrical standards)
IEE	Institution of Electrical Engineers
IOSH	Institution of Occupational Safety and Health
IPC	Integrated polluton control
IQ	Intelligence quotient
IRLR	Industrial relations law report
ISO	International Standards Organisation
ISRS	International Safety Rating System

JHA	Job hazard analysis
JP	Justice of the Peace
JSA	Job Safety Analysis

| KB | King's Bench |
| KISS | Keep it short and simple |

LA	Local Authority
LEL	Lower explosive limit
$L_{EP.d}$	Daily personal noise exposure
LEV	Local exhaust ventilation
LJ	Lord Justice
LOLER	Lifting Operations and Lifting Equipment Regulations 1998
LPG	Liquefied petroleum gas
LR	Lifts Regulations 1997
lv/hv	Low volume high velocity (extract system)

mcb	Miniature circuit breaker
MEL	Maximum exposure limit
MHOR	The Manual Handling Operations Regulations 1992
MHSW	The Management of Health and Safety at Work Regulations 1992
MOSAR	Method organised for systematic analysis of risk
MPL	Maximum potential loss
M.R.	Master of the Rolls

NC	Noise criteria (curves)
NDT	Non-destructive testing
NEBOSH	National Examination Board in Occupational Safety and Health
NI	Northern Ireland Law Report
NIHH	The Notification of Installations Handling Hazardous Substances Regulations 1982
NIJB	Northern Ireland Judgements Bulletin (Bluebook)
NLJ	Northern Ireland Legal Journal
NONS	The Notification of New Substances Regulations 1993
npf	Nominal protection factor
NR	Noise rating (curves)
NRA	National Rivers Authority
NRPB	National Radiological Protection Board
NZLR	New Zealand Law Report

OJ	Official journal of the European Community
OECD	Organisation for Economic Development and Co-operation
OES	Occupational exposure standard
OFT	Office of Fair Trading
OR	Operational research

P4SR	Predicted 4 hour sweat rate
Pa	Pascal
PAT	Portable appliance tester
PC	Personal computer
PCB	Polychlorinated biphenyl
PHA	Preliminary hazard analysis
PMNL	Polymorphonuclear leukocyte
PPE	Personal protective equipment
ppm	Parts per million
ptfe	Polytetrafluoroethylene
PTW	Permit to work
PUWER	The Provision and Use of Work Equipment Regulations 1998
PVC	Polyvinyl chloride
QA	Quality assurance
QB	Queen's Bench
QMV	Qualifies majority voting
QUENSH	Quality, environment, safety and health management systems
r.	A clause or regulations of a Regulation
RAD	Reactive airways dysfunction
RCD	Residual cirrent device
RGN	Registered general nurse
RIDDOR	The Reporting of Injuries, Diseases and Dangerous Occurrences Regulations 1995
RM	Resident magistrate
RoSPA	Royal Society for the Prevention of Accidents
RPA	Radiation protection adviser
RPE	Respiratory protective equipment
RPS	Radiation protection supervisor
RR	Risk rating
RRP	Recommended retail price
RSI	Repetitive strain injury
s.	Clause or section of an Act
SAFed	Safety Assessment Federation
SC	Sessions case (in Scotland)
Sen	Sensitizer
SEN	State enrolled nurse
SIESO	Society of Industrial Emergency Services Officers
Sk	Skin (absorption of hazardous substances)
SLT	Scottish Law Times
SMSR	The Supply of Machinery (Safety) Regulations 1992
SPL	Sound pressure level
SRI	Sound reduction index
SRN	State registered nurse
SRSC	The Safety Representatives and Safety Committee Regulations 1977

SSP	Statutory sick pay
Sv	Sievert
SWL	Safe working load
SWORD	Surveillance of work related respiratory diseases
TLV	Threshold Limit Value
TUC	Trades Union Congress
TWA	Time Weighted Average
UEL	Upper explosive limit
UK	United Kingdom
UKAEA	United Kingdom Atomic Energy Authority
UKAS	United Kingdom Accreditation Service
v.	versus
VAT	Value added tax
VCM .	Vinyl chloride monomer
vdt	Visual display terminal
VWF	Vibration white finger
WATCH	Working Group on the Assessment of Toxic Chemicals
WBGT	Wet bulb globe temperature
WDA	Waste Disposal Authority
WHSWR	The Workplace (Health, Safety and Welfare) Regulations 1992
WLL	Working load limit
WLR	Weekly Law Report
WRULD	Work related upper limb disorder
ZPP	Zinc protoporphyrin

Organisations providing safety information

Institution of Occupational Safety and Health, The Grange, Highfield Drive, Wigston, Leicester LE18 1NN 0116 257 3100

National Examination Board in Occupation Safety and Health, NEBOSH, 5 Dominus Way, Meridian Business Park, Leicester LE3 2QW 0116 263 4700 Fax 0116 282 4000

Royal Society for the Prevention of Accidents, Edgbaston Park, 353 Bristol Road, Birmingham B5 7ST 0121 248 2222

British Standards Institution, 389 Chiswick High Road, London W4 4AL 0181 996 9000

Health and Safety Commission, Rose Court, 2 Southwark Bridge, London SE1 9HS 0171 717 6600

Health and Safety Executive, Enquiry Point, Magnum House, Stanley Precinct, Trinity Road, Bootle, Liverpool L20 3QY 0151 951 4000 or any local offices of the HSE

HSE Books, PO Box 1999, Sudbury, Suffolk CO10 6FS 01787 881165

Employment Medical Advisory Service, Daniel House, Trinity Road, Bootle, Liverpool L20 3TW 0151 951 4000

Institution of Fire Engineers, 148 New Walk, Leicester LE1 7QB 0116 255 3654

Medical Commission on Accident Prevention, 35–43 Lincolns Inn Fields, London WC2A 3PN 0171 242 3176

The Asbestos Information Centre Ltd, PO Box 69, Widnes, Cheshire WA8 9GW 0151 420 5866

Chemical Industry Association, King's Building, Smith Square, London SW1P 3JJ 0171 834 3399

Institute of Materials Handling, Cranfield Institute of Technology, Cranfield, Bedford MK43 0AL 01234 750662

National Institute for Occupational Safety and Health, 5600 Fishers Lane, Rockville, Maryland, 20852, USA

Noise Abatement Society, PO Box 518, Eynsford, Dartford, Kent DA4 0LL 01322 862789

Home Office, 50 Queen Anne's Gate, London SW1A 9AT 0171 273 4000
  Fire Services Inspectorate, Horseferry House, Dean Ryle Street, London
  SW1P 2AW 0171 217 8728

Department of Trade and Industry: all Departments on 0171 215 5000
  Consumer Safety Unit, General Product Safety: 1 Victoria Street,
  London SW1H 0ET
  Gas and Electrical Appliances: 151 Buckingham Palace Road, London
  SW1W 9SS
  Manufacturing Technology Division, 151 Buckingham Palace Road,
  London SW1W 9SS

Department of Transport
  Road and Vehicle Safety Directorate, Great Minster House, 76 Marsham
  Street, London SW1P 4DR 0171 271 5000

Advisory, Conciliation and Arbitration Service (ACAS), Brandon House,
180 Borough High Street, London SE1 1LW 0171 396 5100

Health Education Authority, Hamilton House, Mabledon Place, London
WC1 0171 383 3833

National Radiological Protection Board (NRPB), Harwell, Didcot, Oxford-
shire OX11 0RQ 01235 831600

Northern Ireland Office
  Health and Safety Inspectorate, 83 Ladas Drive, Belfast BT6 9FJ 01232
  701444
  Agricultural Inspectorate, Dundonald House, Upper Newtownwards
  Road, Belfast BT4 3SU 01232 65011 ext: 604
  Employment Medical Advisory Service, Royston House, 34 Upper
  Queen Street, Belfast BT1 6FX 01232 233045 ext: 58

Commission of the European Communities, Information Office, 8
Storey's Gate, London SW1P 3AT 0171 222 8122

British Safety Council, National Safety Centre, Chancellor's Road,
London W6 9RS 0171 741 1231/2371

Confederation of British Industry, Centre Point, 103 New Oxford Street,
London WC1A 1DU 0171 379 7400

Safety Assessment Federation (SAFed), Nutmeg House, 60 Gainsford
Street, Butler's Wharf, London SE1 2NY 0171 403 0987

Railway Inspectorate, Rose Court, 2 Southwark Bridge, London SE1 9HS
0171 717 6630

Inspectorate of Pollution, Romney House, 43 Marsham Street, London
SW1P 3PY 0171 276 8083

Back Pain and Spinal Injuries Association, Brockley Hill, Stanmore,
Middlesex 0181 954 0701

Appendix 5

List of Statutes, Regulations and Orders

Note: This list covers all four volumes of the Series. Entries and page numbers in bold are entries specific to this volume. The prefix number indicates the volume and the suffix number the page number in that volume.

List of Cases

Note: This list covers all four volumes of the Series. Entries and page numbers in bold are entries specific to this volume. The prefix number indicates the volume and the suffix number the page in that volume.

Cadbury Ltd v. Halliday (1975) 2 All ER 226, *1.115*
Carlill v. Carbolic Smoke Ball Co. (1893) 1 QB 256, *1.78*
Close v. Steel Company of Wales (1962) AC 367, *1.38*
Cunningham v. Reading Football Club (1991) *The Independent*, 20 March 1991, *1.40*

Darbishire v. Warren (1963) 3 All ER 310, *1.80*
Davie v. New Merton Board Mills Ltd (1959) 1 All ER 67, *1.156*
Davies v. De Havilland Aircraft Company Ltd (1950) 2 All ER 582, *1.138*
Director General of Fair Trading v. Tyler Barrett and Co. Ltd (1 July 1997, unreported), *1.127*
Dixons Ltd v. Barnett (1988) BTLC 311, *1.116*
Donoghue (McAlister) v. Stevenson (1932) All ER Reprints 1, *1.143*
Donoghue v. Stevenson (1932) AC 562, 38, 149, *1.152*
Dunlop Pneumatic Tyre Co. Ltd v. Selfridge & Co. Ltd (1915) AC 847, *1.78*

East Lindsay District Council v. Daubny (1977) IRLR 181, *1.104*
Edwards v. National Coal Board (1949) 1 KB 704, (1949) 1 All ER 743, 2.157, *4.198*
European Court of Justice cases
  C382/92 Safeguarding of employee rights in the event of transfer of undertakings, Celex no. 692JO382, EU Luxembourg (1992), *1.98*
  C383/92 Collective Redundancies, Celex no. 692JO383, EU Luxembourg (1992), *1.98*

Factortame Ltd No. 5, Times Law Reports, 28 April 1998, *1.26*
Fenton v. Thorley & Co. Ltd (1903) AC 443, 2.4
Fitch v. Dewes (1921) 2 AC 158, *1.80*
Fitzgerald v. Lane (1988) 3 WLR 356, *1.151*, *1.157*
Fletcher Construction Co. Ltd v. Webster (1948) NZLR 514, *1.150*
Frost v. John Summers and Son Ltd (1955) 1 All ER 870, *1.137*

General Cleaning Contractors Ltd v. Christmas (1952) 2 All ER 1110, *1.138*
General Cleaning Contractors Ltd v. Christmas (1953) AC 180, *1.154*
George Mitchell (Chesterhall) Ltd v. Finney Lock Seeds Ltd (1983) 2 All ER 737, *1.85*
Global Marketing Europe (UK) Ltd v. Berkshire County Council Department of Trading Standards (1995) Crim LR 431, *1.117*

Hadley v. Baxendale (1854) 9 Exch. 341, *1.80*
Heal v. Garringtons, unreported, 26 May 1982, *1.141*
Hedley Byrne & Co. Ltd v. Heller & Partners Ltd (1964) AC 463, *1.17*, *1.38*
Henderson v. Henry E. Jenkins & Sons (1969) 3 All ER 756, *1.138*
Henry Kendall & Sons v. William Lillico & Sons Ltd (1968) 2 All ER 444, *1.84*
Hicks v. Sullam (1983) MR 122, *1.116*

Taylor v. Alidair Ltd (1978) IRLR 82, 104–5

Tesco Supermarkets v. Nattrass (1972) AC 153, *1.116*

Thompson v. Smiths Ship Repairers (North Shields) Ltd (1984) 1 All ER 881, *1.140*

Toys R Us v. Gloucestershire County Council (1994) 158 JP 338, *1.118*

Treganowan v. Robert Knee & Co. Ltd (1975) IRLR 247; (1975) ICR 405, *1.109*

Vandyke v. Fender (1970) 2 All ER 335, *1.137*

Victoria Laundry (Windsor) Ltd v. Newman Industries Ltd (1949) 2 KB 528, *1.80*

Walton v. British Leyland (UK) Ltd (12 July 1978, unreported), *1.124*

Ward v. Tesco Stores (1976) 1 All ER 219, *1.139*

Waugh v. British Railways Board (1979) 2 All ER 1169, 20, *1.144*

Wheat v. Lacon & Co. Ltd (1966) 1 All ER 35, *1.87*

Williams v. Compair Maxam Ltd (1982) ICR 800, *1.108*

Wilsher v. Essex Health Authority (1989) 2 WLR 557, *1.151*

Wilson v. Rickett, Cockerell & Co. Ltd (1954) 1 All ER 868, *1.84*

Wilsons and Clyde Coal Co. Ltd v. English (1938) AC 57 (HL), 2.18, 2.157

Wing Ltd v. Ellis (1985) AC 272, *1.118*

Young v. Bristol Aeroplane Company Ltd (1994) 2 All ER 293, *1.133*

Series Index

Note: This list covers all four volumes of the Series. Entries and page numbers in bold are entries specific to this volume. The prefix number indicates the volume and the suffix number the page in that volume.